KEY CONCEPTS IN PLANNING

The *Key Concepts in Human Geography* series is intended to provide a set of companion texts for the core fields of the discipline. To date, students and academics have been relatively poorly served with regard to detailed discussions of the key concepts that geographers use to think about and understand the world. Dictionary entries are usually terse and restricted in their depth of explanation. Student textbooks tend to provide broad overviews of particular topics or the philosophy of Human Geography, but rarely provide a detailed overview of particular concepts, their premises, development over time and empirical use. Research monographs most often focus on particular issues and a limited number of concepts at a very advanced level, so do not offer an expansive and accessible overview of the variety of concepts in use within a subdiscipline.

The *Key Concepts in Human Geography* series seeks to fill this gap, providing detailed description and discussion of the concepts that are at the heart of theoretical and empirical research in contemporary Human Geography. Each book consists of an introductory chapter that outlines the major conceptual developments over time along with approximately twenty-five entries on the core concepts that constitute the theoretical toolkit of geographers working within a specific subdiscipline. Each entry provides a detailed explanation of the concept, outlining contested definitions and approaches, the evolution of how the concept has been used to understand particular geographic phenomena, and suggested further reading. In so doing, each book constitutes an invaluable companion guide to geographers grappling with how to research, understand and explain the world we inhabit.

Rob Kitchin
Series Editor

KEY CONCEPTS IN
PLANNING

GAVIN PARKER
AND JOE DOAK

Los Angeles | London | New Delhi
Singapore | Washington DC

Los Angeles | London | New Delhi
Singapore | Washington DC

SAGE Publications Ltd
1 Oliver's Yard
55 City Road
London EC1Y 1SP

SAGE Publications Inc.
2455 Teller Road
Thousand Oaks, California 91320

SAGE Publications India Pvt Ltd
B 1/I 1 Mohan Cooperative Industrial Area
Mathura Road
New Delhi 110 044

SAGE Publications Asia-Pacific Pte Ltd
3 Church Street
#10-04 Samsung Hub
Singapore 049483

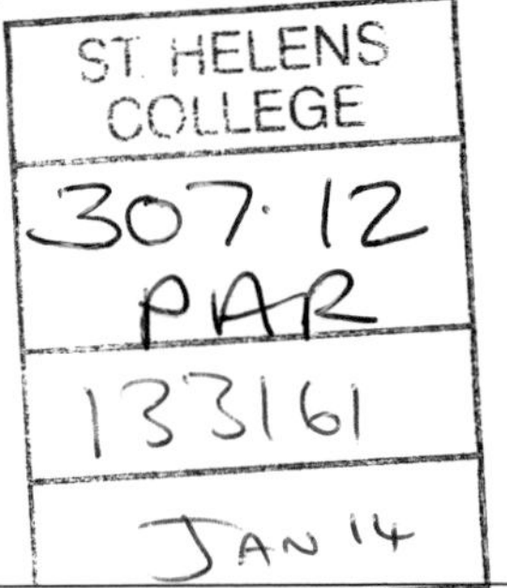

Editor: Robert Rojek
Editorial assistant: Alana Clogan
Production editor: Katherine Haw
Copyeditor: Lotika Singha
Proofreader: Helen Fairlie
Marketing manager: Michael Ainsley
Cover design: Wendy Scott
Typeset by: C&M Digitals (P) Ltd
Printed by: Replika Press Pvt. Ltd

First published 2012

Library of Congress Control Number: 2012935022

British Library Cataloguing in Publication data

A catalogue record for this book is available from the British Library

ISBN 978-1-84787-076-6
ISBN 978-1-84787-077-3 (pbk)

For Sandra, Gemma, Becky and Alex: the past, present and future flows through you.

CONTENTS

ABOUT THE AUTHORS

Gavin Parker is Professor of Planning Studies at the University of Reading, UK.

Joe Doak is Senior Lecturer in Urban Planning and Development at the University of Reading, UK.

ACKNOWLEDGEMENTS

Thanks to Sarah-Jane Boyd and Katherine Haw at Sage for their patience, and to Lotika Singha for her tolerant copy editing. We would also like to thank our colleagues and students at the University of Reading for their support in writing this book. Meiko and Mandy suffered and supported us in getting this produced amongst the interstices of life.

1 INTRODUCTION

Why this Book?

Planning as a subject for study is one of the most multidisciplinary: practice and theory cut across and draw on a breadth of social sciences and other influences. It is an integrative concern, an endlessly dynamic and interesting field for study and reflection, Moreover the aims, activities and implications of planning affect us all. The concepts included here and the way that they are applied have been shaped and reshaped by policy makers and practitioners, by commentators and academics within planning and across the contributory disciplines. These ideas and labels are actively shaping practice and are shaped by practice – they are fluid and open to appropriation. All of these concepts therefore reflect and can be used to understand planning activity in different ways. In selecting these key concepts we have included ideas that provide a basis for understanding planning and the factors that shape the relationship between society and environment. Planning policies tend to reflect social choices about resource use and the organisation of the built and natural environment.

One of the founders of modern sociology, Emile Durkheim, contended that a concept was a collective and abstracted representation taken from the flow of time and space (cited in Urry, 2000b: 26). While this view may sound rather daunting, it intimates that not only are concepts contingent and imperfect, but also that they are crucial in communicating ideas. Dictionary definitions indicate that a concept can be a new idea or that which encapsulates some otherwise abstract idea. Concepts can also be seen as accessible or 'compressed packages' of theory and practice. They are constantly being traded and modified, used and applied by different people for a variety of reasons. As such it is useful to 'freeze' the process and unpack some of the key ideas that have influenced the planning community over the past half-century or more. Presenting them here is important both to ensure that they are understood and that their contingent nature is revealed. By doing this, one can 'place' oneself and assess specific practices within the planning field in a more considered way.

This task is part of a wider need to understand theory and research methods that have long been established as core elements of planning education (e.g. Royal Town Planning Institute (RTPI), 2011). This is largely because analytical skills and an ability to critique research and deconstruct practice are widely recognised as important tools for planners. Often there is a perceived divide between theory and practice, and significant difficulties exist in trying to bridge this gap. Many planning books concentrate on the facts, rules and institutional arrangements involved in planning without contextualising or putting these into a conceptual framework. Conversely, students often feel that courses covering theory and methods can be rather abstract and disconnected from practice. This may also be because theory texts tend to consider theory in paradigmatic or conceptual silos, are written in inaccessible language, or are otherwise seemingly detached from practice. In our view, this situation can be alleviated by a clearer explanation of the key concepts that planners use or that otherwise influence action.

Beyond the standard theory and research methods texts there are few books available for students of planning that systematically explore the ideas and structuring concepts that planners may need in order to understand and apply theory and to assist in reflecting on practice. The idea is not to suggest that 'planning' can be broken down and covered comprehensively through the key concepts here; rather, these ideas should be indicative and illuminating for people who are entering and operating in policy fields that shape or sustain the relevance of these concepts. These abstractions do form much of the 'operating logic' of planning and a basis for argumentation in planning. As such we hope that this text will act as a useful reference for anyone engaging with planning systems, with development processes or with land use policy, but in particular for students of town planning, human geography and other related fields.

Planners, in the widest sense, need to be able to conceptualise the processes, justifications and conditions of operation that surround them. Rather than being experts or possessing an unmanageable load of 'expert knowledge', our view is that twenty-first century planners need to be able to understand and 'place' the expert (and lay) knowledges of others and then proceed towards open and clear deliberation and decision-making. Part of this open toolkit involves a wide appreciation of ideas, drivers and difficulties associated with planning and the management of environmental change.

In this spirit, a standard introductory text on planning in the UK includes in its preface the remark that: 'It is not easy (or even useful) to define the boundaries of town and country planning ... planning policies are now far broader [than ever before]. Moreover the importance of interrelationships with other spheres of policy, which has long been accepted, is now enshrined in the "spatial planning approach"' (Cullingworth and Nadin, 2006: xix). This sentiment underlines and extends the importance of conceptual awareness as a means of seeing the linkages, overlaps and gaps in spatial planning and indeed with planning practices comparatively. Without an arsenal of concepts that are well understood and forged through debate and honed through reflexive application, planning practice is unlikely to be robust or effective. Linkages can be made more easily across concepts and fields of planning, and an understanding of how issues are framed can be established more readily.

We cannot hope to be comprehensive in our coverage and instead we have included a selection of key concepts and organised them so that they can be read as an entirety, in groups or individually as a reference source. Contributory or overlapping ideas are indicated in each concept chapter. This underlines the way in which concepts are linked and where similar ideas may be discussed using synonyms or other labels.

Concepts 'For' and 'Of' Planning

This book provides a guide to ideas that shape planning practice and research, which are concepts both 'for' and 'of' planning and this is a deliberate choice; to draw in from outside the planning 'discipline'. Some concepts are generally discussed and are quite widely applicable in the social sciences, while others have been developed from the activities and concerns of planning practice. In some cases there may be different meanings or applications in circulation. Every concept expanded on in this book has its own history or 'back-story' that needs to be appreciated to some extent to understand its utility and limitations. In reading these concepts and the related explanations, the reader should be able to reflect on the assumptions that are brought to bear on situations where these concepts are 'in the script'. There are no simple 'answers' or 'off the shelf' solutions or understandings in applying these. Pre-assembled attitudes and experiences, combined with the situation at hand, will substantially shape understanding.

The ongoing process of formulating and negotiating conceptual development is often in tension with practice and empirical knowledge. The idea of praxis as a dialogical engagement between theory and practice is a sound principle, yet is often hard to maintain. Practising planners do need to retain the sensibilities of researchers if they are to have any hope of achieving a dynamic and relevant approach to planning in the coming decades, as do researchers or teachers in relation to practice. While we do not intend to over-theorise and complicate this book, it is important that students understand the terms outlined below. This is not because they appear explicitly in the text but rather, they act as a reminder to absorb the possibilities of each chapter in the light of different ways of viewing the world.

Three principles of knowledge and methodology can be usefully dusted down and employed as a reflexive filter when reading this text:

- The first is the idea of *ontological disposition* – the attitude taken towards knowledge, truth and legitimacy and which therefore highlights 'what' is to be studied. This puts an onus on planners to consider the suppositions or empirical assumptions made by any particular theory, or that are contained in any conceptual explanation.
- The second principle is best encapsulated by the term *epistemological perspective*. That label implies that there are different ways of conceptualising the world and different ways to research it. Individuals are likely to favour particular approaches, readings or parameters that constitute validity. This may involve an explicit 'ruling out' of particular forms of knowledge, inputs or methods of knowledge collection, as a result of particular pedagogic styles, or what could be a result of socialisation processes. This connects to normalisation, and implicit favouring of particular representations of knowledge, 'truth' or 'fact'. This highlights that we tend to rule-out or rule-in different ideas, 'facts' or influences depending on how they may suit our purposes or chime with our pre-existing attitudes. In this way, the use of different research methods, as below, may promote or underpin different epistemological positions.
- Third, *methodological approaches* are different ways of collecting and analysing data (see, for example, Alasuutari, 1995; Bryman,

> 1988; Silverman, 2000). These influence and are instrumental in constructing ontological and epistemological 'realities'. By choosing different techniques and approaches, the planning researcher (in academia or practice) tends to prioritise certain types of data over others. Methods used, knowingly or otherwise, also tend to bring different values and biases into the analysis – which then lead to conclusions and recommendations. Openness and awareness of these methodological predispositions and the strengths and weaknesses of each should be a core attribute of the 'thinking planner'.

Methodological, epistemological, and ontological stances or views recursively affect each other in some way. Another useful technique is to review the concept from the perspective of different interests or parties in any given situation; asking what ontological or epistemological stances are typically adopted by those interests (see also Chapter 9); and how different data and methods may influence argumentation and decision-making in planning.

Planning and Planners: Skills and Understandings for the Twenty-first Century

It is to be hoped that our initial comments are not too off-putting: as mentioned, there is a tendency for people to be apprehensive about studying theory. It can appear daunting or lack a clear application to future careers, practice or to life in general. While we contend that theory is important per se, we also see that linking theories to keywords and ideas can help highlight the utility of theorisation and render ideas, concepts and theories more intelligible and revealing. As such, this is not a theory book but more of an explanation of how and where ideas are formed and used in planning contexts.

Furthermore the demands on planning and planners are seemingly ever widening in terms of the objectives or goals and skills and knowledges required, and to such an extent that planning-related activity is undertaken by a wide range of practitioners, or studied by a variety of disciplines in the academy. As indicated, planning as an activity is more diverse and complex than ever, and planners are being encouraged to adopt a broad and flexible conceptualisation of planning. For example,

Cullingworth and Nadin's (2006) overview of UK planning stands at almost 600 pages (with 11 pages of acronyms!).

What is planning anyway?

So far this chapter has concerned itself with the justification for the content of the following chapters and pointed towards the breadth of the subject or enterprise of planning. The concepts covered here reflect this breadth, but before we embark we feel it is useful to outline loosely 'what planning is', with this theme deepened in Chapter 2.

The growth in planning concerns and activity of different types has in some ways obscured what planning is or might be. Planning may mean different things to different people, and understandings (as well as practices and scope) have changed over time. When approaching the topic it can on some level appear beguilingly simple. The word planning appears rather innocuous on the surface – it is after all an everyday word that can incorporate the quotidian. For others it is heavy with meaning, implying 'organisation' or perhaps centralised control. There is a political dimension that pervades these attitudes. Byrne highlights how 'planning if it is anything, is a way of changing things – a mode of transformation' (2003: 172). This definition places planning as a means of effecting change, but it is also about shaping and speeding or slowing change. As a result, the operation of planning policies and powers is intrinsically politicised and planning has been open to critique from both the left and the right on the political spectrum: both in terms of mediating and ameliorating unfair spatial and economic outcomes and in terms of restricting individual freedom. This contestation has ranged from seeing planning as a means to 'disguise oppression in the language of liberal hope' (Hoch, 1996: 32), through to a justification for collective power to dominate 'rational' markets. This is important given the way that people respond to different plans, planners and policies that are wrapped up in ideological and interest-based concerns. Increasingly across Europe and elsewhere the market has become the engine of economies and of 'development'. Where this is so, planning is seen by some as providing necessary steering equipment, while others perceive it as an (unnecessary or unwanted) interference in the operation of the 'free' market. This view, of course, ignores the fact that all organisations, be they from the 'public' or 'private' sector, engage in forms of planning to 'transform' their activities or environment to achieve certain objectives.

We are concerned here primarily with spatial planning involving the orchestration and management of development and of the interrelationships between people and place and various uses and activities. In this light, Healey (2006) identifies three strands or 'traditions' of planning. The first and possibly the most immediate form which tends to loom large in people's minds is physical planning, which involves management of development so that it is appropriate to its context. This is concerned with physical form and the interrelationships of function. The early British planners such as Ebenezer Howard, Raymond Unwin and, subsequently, Patrick Abercrombie certainly fell into this category, with many post-war planners and local government in the UK concentrating their efforts and resources on masterplanning, modernisation and reconstruction of neighbourhoods and whole towns. Healey succinctly puts it that in the past, planners saw urban problems as tasks to be resolved through intervention: 'the challenge was to find a way of organising activities which was functionally efficient, convenient to all those involved, and aesthetically pleasing as well' (2006: 18). As a result of this motivating aim, the focus was largely on achieving appropriate urban forms (see Ward, 1994). However, this objective has often been frustrated by insufficient regulatory powers and resources, or otherwise a lack of consensus and cooperation over the aims and impacts of planning schemes and policies. Moreover, shifting political, economic, technological and social conditions have played a part in undermining planners' aspirations in this regard; and the legitimacy and expertise of planners as arbiters of such processes has also been called into question (e.g. Davies, 1972; Klosterman, 1985).

Economic planning is the second approach that also has fallen foul of socio-economic change and dominant or conflicting political sensibilities. Such 'planning' seeks to manage and shape the economy at different scales or in different ways or directions. This of course is a task that falls on many and only some might actually recognise or define themselves as 'planners' – more likely economists. Ideals of efficiency and rationality in terms of how resources are used and distributed underpin this form of planning, and as a result it has been criticised as politically motivated and inefficient (for example; Evans, 2004; Webster, 1998). The fortunes of central economic planning in the West have met a similar fate to 'command and control' policies found in more totalitarian regimes, both past and present. One problem has been the difficulty in justifying and implementing the inherently top-down decisions made in the name of efficiency and redistribution. Indeed the rationale for this style of economic planning has been influenced strongly by Marxist critiques of capitalism in aiming to redistribute economic benefits. This was often carried through via the construction and maintenance of welfarist regimes subsequently informed by Keynesian economics. However, such central control proved difficult to orchestrate, was criticised as undemocratic and if anything became socially regressive. Instead much economic planning has emerged as a market-curbing partnership between 'key market actors' and governments, where fiscal tools and policies remain as the main mechanisms for economic control but without outright ownership of production (see Adams et al., 2005). Aspects of economic planning can be seen through national policy statements and regional economic strategies, as well as more directly in the decisions and operation of finance ministries and budget allocations from the supranational to the local scale (see also Chapter 17).

Some planning structures and planners have also been criticised for being unjustifiably ideologically biased, or otherwise narrow, in attempting to objectify knowledge and hold up 'facts' or to propose limited alternatives on which others may base their decisions. This introduces the third form that planning takes: policy analysis and the administration of public policy (Healey, 2006). This is an important role and is largely concerned with the setting and implementation of public policy goals at the national and local scales, but is also increasingly about co-constructing and translating international agendas. It is somewhat beyond the concern with economic performance and production found with economic planning as it is also concerned with the attainment of more specific targets. This aspect of planning then is really about *how* to achieve fair and democratic ways of identifying objectives and then devising policies and programmes to achieve them. There is a second aspect of policy analysis which is concerned with the careful organisation of knowledge to inform planning: this may be seen as the *why* plan dimension – which is discussed in Chapter 2 and is one of the justifications for planning outlined by Klosterman (1985). How such knowledge is collected and interpreted is somewhat problematic and, as already suggested, the way that knowledge and methods are framed and deployed affects policy analysis immensely. According to Faludi (1973), this has meant that policy analysts have often attempted to decontextualise evidence or information from its political and institutional conditions, or from context, and presented technical solutions or options based on 'scientific knowledge'.

These three traditions illustrate the notion that how planning is viewed and framed will determine its extent and the relevance of the concepts. Furthermore, how information and knowledge is bounded and categorised will influence planning practice. For our purposes, planning is a combination of all these types. Any given role or position in planning will involve a mix and a balancing of all of these strands, with emphasis being placed on different aspects, in different areas, at different times. The context of the political conditions found in those areas and times is also likely to shape the practice of planning. Therefore, the appreciation and relative importance of these strands of planning are contingent on the political climate and relative power of different interests in planning. The way that options and policies are formulated through these types of planning requires careful scrutiny, with the concepts and associated ideas contained in the following chapters helping in this process.

Since the 1990s the key policy aim of planning, arguably across the world, but certainly in the UK and much of Europe, has centred around the notion of sustainable development or sustainability, as discussed in Chapter 3. In global terms the impact of several rounds of environmental summit meetings (Rio de Janeiro, Johannesburg, Kyoto, Copenhagen) has served to underpin this agenda. However, the balancing of the economic, environmental and social dimensions of sustainability has involved an uneven and often uneasy amalgamation and reorganisation of the different planning strands outlined above. Although sustainability has been gradually embedded as the touchstone or 'metanarrative' of public policy, and the conceptual framework for much of contemporary policy analysis, it has spilled over to shape economic planning regimes (for example via calls for industrial ecology, environmental management, corporate social responsibility and carbon footprinting and other reduction or mitigation measures) and in physical planning (for example, in the almost obsessive search for the most sustainable urban form (see Breheny, 1992; Jenks, 2005; Jenks and Burgess, 2000)).

Planning and its specialisms

The three traditions conceal numerous fields or specialisms, such as transport planning, urban design, rural planning, conservation planning and waste planning. Planning is so broad that no single planner is likely to be 'expert' across such a range of activities and scales. Therefore, it is increasingly important that planners are aware of, and understand, the unifying concepts that influence the range of planning activity. Indeed if planning is concerned with interrelationships, as suggested above, then this text acts as a kind of translator and 'bridge' that can help to link different parts of their studies or experiences. Moreover, establishing, or at least committing to paper, a set of boundary-crossing concepts for planners to share helps to ensure a degree of common understanding and provides a useful toolkit of ideas.

Understanding where different ideas and expertise fit into the 'real world' or how they may influence the policy aims, or aspirations of different interests, or indeed other ideas that are in circulation, is the next best thing to actually being a polyglot. In one sense this provides the basis of Schön's (1983) 'reflective practitioner' who seeks to critically engage with the concepts that shape their professional practice (see Healey, 2006; also, Murdoch, 2005). For us, the ability to consciously reflect on and act to (re)construct spatial planning practice is a prerequisite skill in a rapidly changing environment and in a politically dynamic context. Planning employers have consistently encouraged universities to allow students to develop and hone analytical skills, and this requires an ability to understand and assess 'real world' situations critically. The Egan review (2004), which investigated skills for planning and related professions in the UK, indicated that key generic skillsets should include: evaluation of alternatives, analytical skills, visioning and creative thinking, and working with other partners or stakeholders. This inherently requires a knowledge and understanding of key concepts and of theory, and the development of a reflexive disposition in order to make connections, inform decisions, and provide a platform for critical and constructive thinking. In this context, the chapters in this book act as horizontal integrators and some will be of close relevance to particular topic areas, while some may be necessary to unlock the content of others.

The Concepts: Range, Selection and Structure of the Chapters

The nature and practice of planning is such that ideas and concepts are drawn into and influence planning from a variety of source disciplines such as geography, sociology, politics and economics. The concepts included here indicate how and where the ideas come from and how they are typically understood or applied. The concepts are chosen either because they are recurring and enduring concepts in planning, or they are, in our view, the main ideas that are closely linked to the integrative key components of planning practice. In recognition of the social constructionist perspective that underpins our approach, we contend that all of the concepts are inherently 'social' in their formation and deployment (see Berger and Luckmann, 1966).

Each of the chapters provides a detailed explanation of the concept, outlining various contested definitions and the evolution of how the concept has been used, as well as links and examples of application or relevance in practice. Furthermore, each concept is associated (at the very start of each chapter) with a family of related sub-concepts or terms that are 'embedded' or linked with the core concept. This illustrates the fact that key concepts do not stand alone, but are part of a tapestry of words and ideas that are used in planning to structure and shape policy and practice. This also means that while concept labels may not necessarily change, the way that concepts are understood or used may well alter over time. The chapters are broadly structured such that the concept is introduced, then the main components and debates are 'unpacked' and explored before the application and use of the concept in relation to planning practice is outlined. A concluding section, briefly drawing out the key themes, is followed by a short note on further reading.

Some concepts (e.g. sustainability, plan, place and community) are considered important enough to have a more extended analysis of their meaning, and these are slightly longer than others. Overall, we have striven to provide not only concise overviews of the key concepts, but also to highlight their breadth and the more apparent interlinkages between them. Other chapters included here are 'Networks' (Chapter 4) in which the linkages between actors and different actants are emphasised; 'Systems and Complexity' (Chapter 5), which picks up on more recent understandings of the interrelations that may be seen between actors and the environment, and indicates how planners may have difficulty in understanding and predicting outcomes or managing change. We also consider implementation (Chapter 7) as a key idea and goal for planning activity in terms of the aim and difficulties in reaching goals. The use of designations (Chapter 8) and the linked concept of hierarchy (Chapter 6) are explained as ways of organising and demarcating space and lines of power. We then move on to think about the different interests (Chapter 9) involved in planning and the commonly cited 'public interest' justification for planning decisions. Negotiation is considered, given the way that many decisions and situations in planning tend to require negotiative responses, as is mobility and accessibility and its consequential implications for planning (Chapters 10 and 11). The concept and implications of rights in planning are covered in Chapter 12, given, in particular, the role of planning in determining and enforcing property rights. Chapter 13 focuses on place, space and sense of place, and discusses these at some length, including aspects of urban design activity. Community (Chapter 14) is then examined. This features as a concern for planners that is linked to questions of sense of place and reflects a wider aim in many countries both to maintain community and to create the conditions for more cohesive societies.

A broad take on the concept of capital (Chapter 15) is explained and unpacked as a means of conceptualising resources which are used, stored, traded and exchanged through policy decisions, development and regulation. An examination of externalities and impacts (Chapter 16) is included as these are often regarded as important justifications for planning regulation. The use and concern for regional and national competitiveness (Chapter 17) is covered, highlighting how planning can aid, but is equally seen sometimes as a brake on, economic activity. The longstanding use of amenity as a rhetorical justification for planning and development control is explained in Chapter 18.

Lastly but by no means least, a consideration of development (Chapter 19) concludes the key concepts selected. This final chapter is important as planning and development tend to go hand-in-hand and significant cross-disciplinary cross-fertilisation has occurred in recent years between the range of built environment disciplines and their professional communities. Planners have been central initiators and shapers of theory in that process, feeding through into the more empirically based paradigms of surveyors, engineers, construction managers and (to a lesser extent) architects. This is reflected in recent debates around the conceptualisation of the development process itself where a number of academics have led the way in deconstructing the various relational webs seen to be involved. Indeed it is through this type of process of reconceptualisation that other established ideas are being opened up and discussed in planning.

Reflecting the 'emergent' nature of planning concepts, we are now seeing modifications to planning policy and practice arising from the need to address climate change. We have not given climate change a separate entry as it clearly relates to the idea of sustainable development (Chapter 3), but its recent rise in importance indicates that the environmental agenda continues to permeate and restructure planning practice and its associated core concepts. We have accommodated this fact specifically and more generally by including cross-references and indicating conceptual linkages where they are appropriate.

This introductory chapter has set out the content and the reason for needing to think conceptually when involved in planning. We also hope that the reader will be able to gain an insight into 'what, why and how' planning has changed and has sought justification over time. Although the dissection of single concepts can provide useful windows into the worlds of planning practice, a more comprehensive and contextualised reading of the concepts delivers a more critical and thoughtful appreciation of the dynamics of planning and interests in planning. We hope that this introductory chapter might help the reader to begin that reflexive process and to see how different ideas, interests and contexts impact on how concepts are deployed, understood and refined.

FURTHER READING

Each chapter has its own suggested readings but there has been a steady stream of core texts on planning theory and their content has evolved over time, reflecting the 'accumulation' and reinterpretation of key concepts. It is interesting to compare the theoretical ideas covered in Bailey (1975) with those in Allmendinger (2009) or trace Patsy Healey's theoretical explorations in Healey (1983 or 1988) with her later work on relational approaches (Healey, 2006). There are a number of standard texts on planning theory, including Taylor (1998), Campbell and Fainstein (2003), and Hillier and Healey (2008). The application of theoretical ideas to practice is a challenging endeavour but has been done well in Forester (1989), Healey et al. (1988), Low (1991) and Flyvbjerg (1996). Some of the sister volumes on key concepts in human geography (Holloway et al., 2009) and urban geography (Latham et al., 2009) are also useful aids. Two more recent texts, Morphet (2010) and Rydin (2010a), add to the resources that describe and delineate the roles and purposes of planning in society.

2 PLAN AND PLANNING

Related terms: strategy; vision; organisation; policy; forecasting; procedural theory; complexity

Introduction

When considering the key concepts involved in planning practice we start with the term planning or plan. Including this as a concept here does, however, run the risk of writing a book within a book but also avoids the opposing danger of overlooking this metaconcept. In Peter Hall's words, including this chapter may amount to a 'total superfluity' in that everything in the book should be in this chapter or under this heading! Despite this warning we have included a consideration of 'planning' that sets out what we mean by the term and how planners and others have pursued land use planning and more recent efforts to develop a 'spatial planning' (Nadin, 2007). We also indicate how and where other authors have attempted to define or outline the idea of planning, and the range of activities and practices involved without becoming overly or too densely theoretical. Recognising the limits of this account, we also flag up where to look next in order to develop further understanding of planning theory, technique and history.

This chapter also embraces linked and commonplace words in the planning lexicon, such as 'strategy' and 'vision'. Our intention here is to uncover the underlying meanings and justifications that are wrapped up in these terms, rather than provide a comprehensive account of planning practice. We invite the reader to reflect on the wide-ranging and diverse practices, justifications and arguments that are found in planning activity. This should act as a good lead into critically assessing the way that planning tends to cohere around dominant ideas, practices and policy trajectories examined in the other key concept chapters.

Defining Planning

The concept of planning is obviously and literally synonymous with our focus in this text and is central to the activities of very many actors involved variously in development, service provision, regeneration, public policy formulation and economic development; which act to influence planners and plans. Some react to planning policy while others are more closely involved in helping to shape and implement plans and policies. Thus we are concerned with both *making* and *applying* policy.

The concept is also relevant both as verb and noun; referring to a process, action or consequence. In its widest sense some form of planning is undertaken by everyone, but despite very wide meanings of the term our main focus is on the orchestrated and knowing attempts made to shape places and form land use plans, or to engage in variations of town planning, as Hall observes:

> planning conventionally means something more limited and precise: it refers to planning with a spatial, or geographical, component, in which the general objective is to provide for a spatial structure of activities (or of land uses) which in some way is better than the pattern that would be achieved without planning. (2002: 3)

For Rydin, there is a recognition of different forms of planning and the roles of the private sector, but it is assumed that the lead is taken by the public sector with the central role being about producing plans and strategies: 'planning is about devising strategies for shaping or protecting the built and natural environment' (2003: 1). Most often this process and practice culminates in some form of spatial representation, that is, a map showing the spatial implications of policies but this is not always so. A formal plan or strategy will clearly set down its aims and objectives or goals in written form. Other definitions, including those from practitioner communities, are reflected in the 2003 report of the European Conference of Ministers responsible for Regional Planning

(CEMAT), which asserts the geographical essence of 'spatial planning' as being somewhat broader and more holistic:

> spatial planning gives geographical expression to the economic, social, cultural and ecological policies of society. It is at the same time a scientific discipline, an administrative technique and a policy developed as an inter-disciplinary and comprehensive approach directed towards a balanced regional development and the physical organisation of space according to an overall strategy. (Council of Europe, 2003: 1)

This view of planning supports a view of planning as a largely regulatory activity that has been dominated by an instrumental rationality or 'scientistic' approach (Relph, 1981), whereby predicting and providing for places has rested upon particular knowledges and rather reductionist methods. Such methods and assumptions have been based around a modernist and collective or utilitarian set of priorities. This has predominantly involved quantitative modelling, the exercise of 'expert' knowledge and the generalised creation of policies and regulations for particular territories. This is a rather limited encapsulation, given that there are many forms and applications or skills required by planners, but it does give an idea of the culture and dominant approach to planning traditionally taken in many advanced economies. In this appreciation, a planner deliberates over future actions – weighing up alternatives and their consequences – the overall aim being to create a plan or set of overlapping strategies to bring about selected objectives. These objectives are often justified in the 'public interest' (see Chapter 9) and the activity of planning is reflected in the process and outcomes of written plans and strategies.

A consideration of the word *plan* set against what planners actually 'do', reveals the term to conceal a diversity of tasks and approaches inherent in different styles or contexts. Cullingworth and Nadin's (2006) overview illustrates the broad range of applications and topic areas associated with land use and spatial planning in the UK (see also Cullingworth, 1999, for an interesting review, and also Glasson and Marshall, 2007). However, it is not only the application of planning to particular fields and to specific topics, but also the process and rationale of planning that requires some consideration here. Planning as *process* or means, and its substantive *aims* or ends, has been under continuous scrutiny, and underpinning justifications and practices rest on several of the other concepts and associated applications explained in this chapter.

Plan and Planning as a Key Concept

If we firstly take a step back and consider the individual and the idea of planning as part of everyday life, then 'planning' processes often remain tacit or unformalised, or remain part of the individual's 'practical consciousness' or repertoire of day-to-day information, reflection and action. This is explained by sociologists such as Giddens (1984) and Bourdieu (1994) (see also Hillier and Rooksby, 2002). From the perspective of the individual, 'planning' may be broadly defined as an intention with forethought, which may or may not be expressed formally. In either case the 'plan' is used as a vehicle to move from a present state, towards the achievement of a range of goals. The plan in its loosest definition is therefore an intended *method* possibly with implied steps or *stages* developed to get from one set of circumstances to another. Often the term strategy is used to express this intent and the means to achieve it, that is, *what* one wants and *how* one means to go about it. A plan is, at its most basic, an expression of an intended course of future action, with planning being the activity that precedes and surrounds that process. These conceptualisations allow for the written or graphical representation of a plan in terms of policies and proposals. The plan is intended to guide the actions of particular groups, communities or wider populations. This definition takes us only so far; there have been different dominant styles of planning over time and which persist in different countries. Some are more focused on design and physical form, some are now more focused on strategic interrelationships and flows between spaces and land uses (i.e. a systems view), while others have moved towards a more integrated form of planning that involves consideration of many different concerns and techniques; often expressed as 'spatial planning' as per the Council of Europe definition above.

We are clearly interested in the method or process of arriving at the plan and in 'planning with particular objectives in mind'. In short this encapsulates and sets up a number of important questions and associated queries relevant in interrogating this key concept:

- *Who is doing the planning?* And is the plan for themselves or for others? How will it impact on others?
- *Why* plan? What are the intended goals or outcomes and the associated objectives?
- *How* to plan? In terms of techniques, methods, assumptions, framing and processes used.

- Under *what conditions* has the plan been formulated? What limitations and constraints prevail?
- Lastly, *who will implement the plan?* What resources, actors, partnerships and conditions will be required for the plan to be realised?

These questions imply consideration of the actors and interests involved in shaping plans and the people, places and activities that are affected or targeted by plans and planning at different scales (see Chapters 6 and 7). We propose that planning requires deconstruction to illustrate and understand its overall purpose in a postmodern context. This importantly involves consideration of the philosophical and methodological underpinnings of spatial planning and the various limitations exposed. We now turn to consider briefly the questions that can help with this.

Who is doing the planning?

As Hall points out, '(w)hether labelled free enterprise or social democratic or socialist, no society on earth today provides goods and services for its people, or schools and colleges for its children, without planning' (2002: 2). Given the broad definitions explored above, we acknowledge this ubiquity of planning processes, yet our focus here is on the arena of spatial planning with its strong state-sponsored dynamics, albeit carried out or led by spatial planners operating in a range of organisational contexts (i.e. public, private or voluntary). This focuses on how planning is carried out in the public interest (Chapter 9) and into an examination of who is involved in the process. Planning, land use planning or spatial planning, has been largely orchestrated by professional planners, using various types of specialist inputs and working with others, including varying degrees of community involvement, to provide needed information or legitimacy into the process. This will typically include a range of data derived from different sources and the data collected will vary depending on the methods used. The planning process will also be shaped through the priorities and limitations set by governments, available resources and the extant legal or procedural frame. Increasingly, questions are asked about the ability and legitimacy of professional planners to carry out planning tasks, and indeed the clear division between the 'planner' and the 'planned' is being blurred, particularly given the different types and scales of planning that exist (and, of themselves, are recognised as fuzzy).

So the question 'who is planning, and for whom?' is more complicated than it may first appear and is also addressed in Chapters 9 and 19, where the role of the planner and the plan is seen as acting to mediate between competing claims and interests. Thus the answer is that planning is being brokered between a wide range of interests and actors. In the UK, national and local politicians have a significant degree of control over both process and policy decisions.

Why plan?

The purposes of planning have been sketched out above and yet the range of specific outcomes or aspirations is diverse. There are several aspects to be highlighted here in terms of normative justifications for planning, such as those outlined by Klosterman (1985). Much planning activity interposes between different interests and their competing goals. Some argue that the 'market' should be the main arbiter (see Lai, 2002; Pennington, 2002), while others indicate that it already is (see Brenner, 2004). In Klosterman's (1985) fourfold justification for planning, the protection of collective interests of the community is foremost, followed by planning acting to improve the knowledge and information base for decision-making and plan preparation. The third justification is having a role in assisting in the protection of disadvantaged and marginalised groups and last is the consideration and mitigation of negative externalities (see Chapter 16 and also Sternberg (1993) on planning in capitalist societies).

The second aspect here is the consideration of planning as *method* (and which crosses over with the 'how to plan' question). This is best explained in terms of efforts to integrate different information sources about future and current needs and pressures on resources. This also concerns the signalling of the direction and spatial distribution of the effects of policy with the intentions of the plan. Statutory plans tend to follow set processes and are prepared with reference to policies prefiguring them in the planning hierarchy (Chapter 6). The initial stages of planning are crucial in terms of goal setting and negotiation across interests or 'satisficing' in terms of evidence gathering and goal-setting (Byrne, 1998; Ham and Hill, 1993; Simon, 1997). One set of problems for statutory planning processes is how to navigate through often competing and conflicting interest positions while adhering to guidelines set from above and in the light of other constraining factors (e.g. legal precedents, resource limitations and interdependencies, and political priorities). Such factors result in ongoing debates about *how* to plan, as described below.

How do planners plan?

This question leads us into methodological questions of how planning is undertaken and the assumptions and purpose of the process of planning. This breaks down to two strands of theory, both of which touch on questions of rationality (Breheny and Hooper, 1985). In short, the first is the substantive (ends-means) side, where the goals of planning are decided and then the best way of achieving them are discussed and constructed. This is based in part on values expressed through governmental policies. The other side of this is the procedural or rational instrumental approach, which reflects a more 'means-ends' balance and where the emphasis is on following particular processes that then suggest or promote particular policies. This tends to follow what has been termed a formal rationality. The latter can be dominant, particularly in market-led economies, where planning can become subservient to the needs of powerful sectional interests and lead to conflict over the legitimacy and purpose of planning decisions. In many countries the reality is that a mix of these two is evident, and the overarching aim of delivering 'sustainable development' is a good example of such a meeting of the instrumental and the substantive.

Regardless of the balance between these two sides, plans and strategies have to be produced somehow and it is desirable that a transparent process is used, and there is long-running disputation over this in many countries. Much has been made about methods of research generally and efforts to confront questions of accountability and inclusivity in planning. Many plans and their formulation will tend to follow standard social science research processes, with a major reliance on aggregated or extrapolated quantitative datasets that has seen the generation of projections and forecasts on which to base decisions (census material being one such source). Typically such large-scale datasets have been compiled at the national level. Local and regional authorities and others gather information on specific issues or add detail to wider questions or requirements to underpin certain aspects of a plan. The reliance on quantitative data has clearly predominated, partly because it has been easier to collect and defend than more qualitative information.

The three classic steps or stages in planning, drawn from the policy-making literature, are *goal formulation*, *identification of objectives* and *target setting*, as part of a goals-hierarchy approach (see Taylor, 1998).

Within these steps there will be processes of evidence gathering, negotiation and the design of implementation tools, and increasingly efforts to develop scenario-based techniques that are aided by advances in technology and software, which guide decisions and encourage participation (e.g. Laurini, 2001). Reforms to planning in the UK have encouraged a more diverse set of inputs and methods in planning but top-down targets and numbers have predominated and substantially formed the parameters of much policy. This process of shifting the style and process of planning in this general direction appears to be a longer-term trajectory, but the principles of operation for planning remain the same across many nations. Rather, it is the emphasis on different stages or tools and the process that vary over time, from place to place, and are the main sources of political dispute, rather than the need to plan and set out agendas.

There is some merit in explaining that the types of method used and constraint placed on the research to generate data are often significant. Information is analysed and interpreted and thence used as evidence to support or otherwise shape a plan. Typically other interests may choose to collect and present their own data, which may conflict, however. Public inquiries are one forum that oversees sectional 'data wars' being played out. The approaches taken and the way in which such efforts are bounded or framed (see Callon, 1998) produce a 'representational affect'. That is to say each reading or use of data and other 'evidence' is partial and subject to variable interpretation and contestation. This is part of the reason that planning is so hotly contested and politicised, and often viewed with suspicion by the public. Moreover, the way that resources and interests are affected by plans in the medium and long term is what induces competing interests to assemble different pieces of evidence that serve their own benefit. This is used to try to shape the plan policies or objectives. Directly linked to this is the simple fact that large amounts of money can be made, or opportunities lost, as outcomes of planning policy and decision making. Equally, significant social and environmental benefits and problems may be created, based on the way that plans and policies are prepared, designed and implemented.

In what conditions are plans made and in which planners operate?

Beyond these questions about the methods deployed in plan formulation there are wider questions about the imperfections of knowledge, interpretation of data and of the conditions in which professional planners work, and thence the questions of implementation explored in Chapter 7. The most obvious issue is the resource limitations that apply to planners and plans. This immediately serves to restrict the way that plans are created. There are other constraining factors to consider, including the need to prepare plans and arrive at decisions within often tight time-frames. These together make for plans that are pragmatic and rely on less than perfect information. The selection of methods and the issue of epistemological positioning are also affected by political and economic boundaries. As such what plans contain is most likely to be bounded by a predetermined scope, perhaps through enabling legislation initially, but from thereon also by resource-based and knowledge-driven boundary setting, interpretation and colonisation of the process by key actors on the ground. The initial purpose of the plan will also dictate how and who is involved in its preparation, as well as the scope of the plan. This will in turn affect the detail and the type of associated powers or resources available to implement the plan (Chapter 7).

The following points are some of the major factors or stages that shape a plan and the process of formulation.

- *Framing* – what is ruled-in and out and by whom. The way that the scope or depth of the analysis used to inform the plan varies. This may be done overtly or less openly and for various reasons as discussed below.
- *Context and history* – what has passed before and what factors have given rise to the need or requirement for the plan. This will tend to influence the view and understandings of what is possible, and includes the influence directly or indirectly of the political environment and wider culture, as well as standing legal principles.
- *Resources and timescales* – these will clearly restrict or enable certain activities and data collection aspects of planning. A longstanding complaint by planners (and about planning) is that they are either hurried or under-resourced or that it takes too long to produce a plan.
- *Political influence* – this is sometimes expressed in terms of policy shaping further up the hierarchy or influence at the local level through direct lobbying by elected local politicians and others.
- *Changing conditions* – unexpected or anticipated change can destabilise or undermine plans and policies. Plans can become obsolescent and this may tempt politicians to opt for broader indicative strategies that may not give the clarity or certainty that some other interests may prefer.

Planning cannot be decoupled from the social, economic and political context in which it operates and to which it responds. The imperfections and biases present in plans, as well as the implementation gap, is a reflection of the field of production. This means that the interrelationships of actors and decisions, legal structures and wider social attitudes produce particular forms of planning and the shape of plans reflect their authorial 'network'; that is to say the people, the conditions and the information that have influenced the plan.

Plans and Planners

It is still the case that in the production of plans, the disposition and training required to draw together data and formulate a plan is very important for planners and others seeking to formulate or influence public policy. These skills are varied, shifting to some extent and contingent on the type of plan being prepared. The plan process inherently involves the use of information and experience to arrive at preferred policy options or policies. In the 1990s, the RTPI developed a competencies list that indicated the generic skills that a planner was expected to possess. The list was supposed to provide guidance on what should be expected of planners at all levels in terms of management knowledge and skills and how best to achieve them. The 15 RTPI (1994) general competencies were:

1. well-developed political skills;
2. strategic management skills;
3. decision-making skills;
4. well-developed negotiation skills;
5. intellect;
6. personal integrity and flexibility;
7. well-developed people management and relationship skills;
8. well-developed communication skills;
9. well-developed influencing skills;
10. results oriented with a drive for achievement;
11. operational management skills;
12. change management skills;
13. self and stress management skills;
14. well-developed analysis and problem-solving skills; and
15. business and commercial skills.

These are interesting and wide-ranging skills, but quite general in terms of plans and plan-making. This reflects the position of the planner as a professional requiring generic skills as well as more specialist or technical skills. As such there are other core learning competencies that professional planners are expected to develop and a separate list has been prepared, and is periodically modified, by the RTPI. For example, these include understanding design principles, research and research methods, the ability to engage with stakeholders, theoretical principles and the political environment – all of which are clearly important (see RTPI, 2011, for the full list of 13 core competencies produced in 2011).

There are numerous other influences and considerations that shape plan making. In terms of town or spatial planning, a course of intended or preferred actions usually relates to a particular bounded space or class of activities, for example a plan for the economic development of a district, or for the management of waste in another area. Each will require different knowledges and data and may require a different process – either due to set or prescribed procedure, or as a reflection of the topics and interests involved. A plan will also affect numerous groups and interests, many of which may be lobbying vigorously and advocating competing ways to proceed towards differing goals or objectives.

Given the above commentary, the idea of planning and the practice of professional planners has broadened, such that many 'planners' will not readily see that they either 'make plans' or act in the ways so far discussed, particularly perhaps those who are involved with development management, and who will be largely concerned with the implementation of planning policy. Indeed those planners working in the private sector will tend to regard plans and draft plans and policies not only as a guide for action, but also positions to be challenged and (re)interpreted. Such planners can only be regarded as plan makers in the sense that they may be involved in lobbying to shape the plan when it is being drafted and in testing its robustness through ongoing challenge and on a case-by-case basis. However, the fundamental purpose of planning embraces the idea that society should anticipate need, risk and priorities in social, environmental and economic terms – the so-called 'planning triangle' (see Chapter 3) and evaluate alternative policy options that correspond to stated policy goals. The 'plan' therefore becomes the crystallisation of this process and is posited by those authorising it as the (best possible) answer to the identified need at that time.

Planning Paradigms and Justifications

The somewhat bland term of 'planning' needs to be revealed as a conflation and reduction of complex and diverse psychological, spatial and social phenomena. Further scrutiny of this somewhat opaque or uninformative label is required, as hidden in the term planning are different methods and techniques variously involved in the process of planning, as well as a divergent set of dispositions or epistemological approaches that can be discerned in planning practice. As mentioned above, this translates into *how* plans are put together and *what* plans contain (and conversely what they may omit).

Different countries have their own legal systems and structures within which planning operates and many states also have explicit legislation that governs the type, purpose and even the timing of plans. The cultures that have developed around planning processes also shape what is ruled-in and out of consideration and what is considered more or less weighty in planning argument. An important part of this will be the extent to which plans are scrutinised and opportunities for public involvement are afforded. Furthermore, some plans are merely indicative and lack enforceable powers, while other planning systems have more regulatory teeth and an even more authoritarian history (see, for example, Yiftachel, 1998).

In the UK, the overarching aim is sustainability and sustainable development, and this has been referred to as a 'meta-narrative' to mobilise planners and other stakeholders (Meadowcroft, 2000), and which aims to shape economic, environmental and social conditions. The broad sustainability framework has also been a driver behind the movement towards a 'spatial planning' approach, in which greater holistic thinking, integrated action and strategic management has been emphasised; the goal being to orchestrate activity and operate within the limits of our resources. Some are also arguing that there is a fourth arm or 'pillar' to sustainability; that is, the cultural dimension (see Hawkes, 2001) that planning needs to more fully understand and factor into decision making (see also Chapters 13 and 18). This reflects a concern that local cultures and traditions are not lost as part of a wider heritage consideration that is seen to reinforce community and quality-of-life questions.

This, however, does not exhaust the range of planning motivations or approaches and this breadth of scope remains a key feature and a strength of planning thought and activity. Indeed the constant debates about the purpose, scope and effectiveness of planning mean that there are multiple discourses attempting to shape planning systems towards different institutional arrangements, policy ends and 'styles' of operation. Brindley et al. (1989) reported on three such styles that evolved and overlapped in the UK (and further afield) during the late twentieth century: trend planning; popular planning; and leverage planning. Each of these rose (and fell) in influence over a period of 20 years and continue to have impact as spatial planning is shaped and reshaped. Studies such as those by Brindley and colleagues illustrate how the dominant paradigms and justifications of planning are forever being made and remade in response to the wider forces of which they are a part.

The Process of Planning

The concept of planning is crucially wrapped up in the organisation of future action based on information and knowledge about the past and present state, allied with projected data using demographic and other statistical information. There are typical steps involved in formal planning that can be mentioned here. In the UK, the planning system has been organised in the past on the basis of a so-called 'predict and provide' model, where forecasting controlled the planning process in a top-down process (i.e. the *trend planning* alluded to by Brindley et al. (1989) above). This has shifted to a modified approach known as 'plan, monitor, manage' to reflect the variability and mutability of conditions in which plans and planning exist. The stages which are hidden in the 'plan' stage, or prior to this, are crucial. As a result, the above shaping of the plan in terms of scope and the process of planning are also closely correlated to the methods used and attitudes of the people who instigate and validate the plan. The selection and use of data are important in this and will depend to some extent on the author's conscious or subconscious epistemological position (e.g. positivist, post-positivist), as well as the possibility of a dominant political or ideological stance, which also affects or biases the process. These understandings and influences impact on methods and analysis with consequential impacts on the final plan.

Standard principles of policy formulation indicate clear stages or idealised forms which contrast with what is often both a messy and opaque process (see Ham and Hill, 1993). Simple, stepped descriptions obscure how the plan is first assembled, how and what is actually monitored, and the ability or powers available to effectively manage the plan's implementation or its effects (intended or otherwise). Cullingworth and Nadin (2006) provide a description of the statutory planning process in the UK, including such stages and checks, and Rydin's text (2003) also explains the development of planning in the UK. However, all of the previously discussed factors need to be accounted for when assessing a planning process (see also Flyvbjerg, 1998, for an account of the power relations present in planning processes), which can be anything other than logical, linear and comprehensive.

Some of these ideas about the planning process have been grouped together into 'procedural theories' of planning (see, for example, Chapter 3 in Allmendinger, 2009; or Chapter 4 in Taylor, 1998), which conceptualise planning as a form of system management or, alternatively, as disjointed and with incremental decision-making. Combining these views into a 'mixed scanning' approach during the 1970s still did not fully capture the complexity of the systemic interrelationships and contexts within which planning processes operate. More recent understandings have acknowledged this (see Chapters 4 and 5 for more detail on this 'relational' perspective).

Conclusion

The idea of planning has been discussed at length with different styles and underpinning logics being provided and critiqued. Such debates have given rise to contestation about the role, legitimacy and utility of planning. The increasingly contingent and complex environment for planning activity has made predicting and imposing plans and policies problematic, both in technical terms and in political and social terms as societies become more diverse, informed and fluid.

The integration of different plans and policies at different levels and across different areas of concern, that is, vertical and horizontal integration, presents another set of challenges for planners and planning policy 'cycles' that may not fit together particularly well. Our concluding comment about planning as a concept and activity is that effective planning must be understood and engaged in as a collective or shared aspiration for the future. For too long planning has been seen as a state-led process of utilitarian management and has been a poorly funded and opaque priority-setting exercise. In the future, new forms of planning are likely to require more open and collaborative elements that both recognise the fuzziness of plan inputs (see de Roo and Porter, 2007) and respond to competing claims to legitimacy and rightness of action.

FURTHER READING

The scale and scope of planning may be appreciated by looking at Cullingworth and Nadin (2006) in terms of the relevant topic areas and activities that planners engage in, albeit from a UK perspective. Hall's (2002) introductory text on planning, particularly the initial chapters, is useful to see how spatial planning has been conceived and practised over time as well, as is the same author's international examination of planning in *Cities of Tomorrow* (Hall, 2000). Many past and current core planning texts (e.g. Greed, 1996; Hall, 2002; Hall and Tewdwr-Jones, 2011; Ratcliffe, 1974; Rydin, 2003) include introductory chapters that review the 'defining elements' of plans and planning, and it is interesting to see the different characteristics and emphases that these authors present. In terms of the underlying epistemological stances and underpinnings of planning and planning theory, Taylor (1998) or Allmendinger (2009) will assist in deepening understanding of how the worldview adopted by the planner can markedly affect the framing of issues or inclusion of different data in plans, and therefore the conclusions and policies that result. Equally classic essays discussing the basis and difficulties experienced by planning can be found in Klosterman (1985), Sternberg (1993) and Friedmann (1998).

Healey (2005) provides an account of how planning might shift towards a position that better reflects the diversity of (post)modern society and the reasons why a more inclusive and deliberative approach to planning may be desirable, and Bruton's *Spirit and Purpose of Planning* (1984) is useful still to indicate how and what has underlain the planning tradition in the UK and elsewhere and is similar in essence to the more recent text by Rydin (2010a). Recourse to looking at the styles and process of planning in different countries will also add value. Key texts for different countries are best left to those living and working in those countries, but some good overviews are available, such as Sanyal (2005).

3 SUSTAINABILITY AND SUSTAINABLE DEVELOPMENT

Related terms: climate change; ecosystems; resource efficiency; greening development; natural capital; environmental impacts; eco-modernisation; carbon reduction

Introduction

Many of the features constituting the broad notion of sustainability have been longstanding and central ideas in planning. The label itself has a relatively recent heritage and sustainability has become the touchstone or explicit primary aim for planning policy and practice in many countries over the past two decades. There has been much attention paid to how planning may aid the promotion of sustainability and sustainable development and vice versa (see Meadowcroft, 2000; Rydin, 2010b). This chapter focuses on providing an understanding of what the term has involved and why it is seen as so important in planning. Despite prolonged and widespread effort, however, sustainability still remains a rather ill-defined and contested concept, and for this and the preceding reasons, it is an obvious choice for early inclusion in this volume.

At its most basic, the word sustainability infers that desirable activity should be able to continue long term or indefinitely, otherwise there is a danger of resource depletion or other environmental impacts that may be difficult or impossible to replace or restore. This general definition applies an anthropocentric perspective with its emphasis on human need. Yet as we will see, the idea covers, and is being used to do service for, a wide range of reasons and it justifies numerous actions at different scales. The associated term 'sustainable development' was created to encompass a set of ideas about the way that human beings could (or should?) live their lives in relation to other human beings, and with regard to the physical environment now and into the future. Although those ideas were created on the basis of people's experience of living with each other and the physical world, the term sustainable development is a socially constructed one – it is given meaning by society. Furthermore, once a term like this comes into existence, it is often deployed and re-created to suit prevailing conditions and attitudes. As such, it suffers from a similar condition that many 'big ideas' do; that it is open to use and abuse, with selective or partial application and with ongoing re-construction, shaped by planners and others. This makes it important to understand why sustainability is seen as being so important and yet so contested, and why it is increasingly dominant as a normative justification or 'meta-narrative' for much public policy (Campbell, 2006; Meadowcroft, 2000; Rydin, 2010b).

Sustainability and Past Efforts to Plan

Spatial or land use planning plays a key role in shaping and directing resource use and in the resulting impacts of carbon emissions (among other externalities – see Chapter 16). Historically, state-led land use planning has used a number of key justifications for intervention in land and development processes and, although the term sustainability is a relatively recent one, the attempt to rationalise and maximise the efficiency of resources and land use has been central to planning theory and practice since its formal foundations in the nineteenth century. Similarly the now widespread threefold social, economic and environmental dimensions of sustainability were loosely presaged by Patrick Geddes' (1915) 'folk-work-place' maxim and much of the thinking underlying the Garden Cities movement just prior to Geddes. These early theorisations highlighted the need to consider the relationships between place (for this, read: environment), folk (read: the social dimension) and work (read: the economic element), and more specifically the needs of industry and of people, while also appreciating and understanding the ecological dimension of sustainable development.

Understandings of this balance and awareness of the interrelationships between the social, environmental and economic have long been argued out between different interests. The idea of measuring impacts and tailoring behaviour accordingly is simply expressed in the 'triple bottom line' (TBL) (Elkington, 1994) approach, as a form of full cost accounting. TBL accounting is where the range of social, ecological and economic impacts are measured and understood before and during all activities. As Meadows explained in 1991, this also helps reveal to all interests what the full implications of our decisions are:

> The world works a little better any time we manage to make the invisible visible, embed real costs into prices, and impose the consequences of decision-making upon those who make the decisions. (1991: n.p.)

Measurement in this way has some obvious repercussions for thinking and gathering evidence about the impacts of new development and for informing planners in relation to spatial planning priorities. Moreover, societal responses to climate change are increasingly requiring better understanding and metric evidence on which to base decisions, often accompanied by forms of systems thinking and the assessment of feedback (see Chapter 5).

Indeed even further back into history it was understood that humans needed to live within the limits of available resources, and many societies developed management and stewardship regimes that ensured long-term survival (see Ostrom, 1990, 2003). As such, the concept of sustainability and the explicit aims contained in models of sustainable development have been embedded in planning for some time and have gained in importance as the world has become more industrialised, more mobile and has used resources more intensively. While conditions have changed and the exact terms used may have altered, contemporary planning practice has found it relatively easy to adapt to recent calls for greater environmental awareness in resource decision-making. This is one reason why many governments have recognised spatial planning as essential in helping to organise and deliver sustainable development.

Decades ago elements of this concept could be found neatly tucked away in the planners' 'balanced' approach to urban development. Indeed Howard's Garden Cities and the later New Towns can be seen as prototype, self-contained 'sustainable settlements'. However, much of the utopian planning thought of the early twentieth century was at odds with prevailing political will, the power of sectional interests or a dominant economic model that prioritised growth and which has not been able to fully reconcile development with its impacts and resource use over time.

Planning efforts have also long targeted other aspects of 'sustainable development'. For example, designations (see Chapter 8) for national parks, green space and historic monuments led the way in protecting valuable environmental and recreational assets in both town and countryside. Overall regional level planning has tried to provide for the long-term needs of the population in a spatially equitable way – although it is arguable if this was as socially equitable or politically sensitive as it could have been. More specifically, years of policy and practice have orchestrated environmental improvements in the inner cities, on urban fringes, in ex-mining areas and the old industrial districts. Through these processes, issues of environmental quality, amenity (Chapter 18) and resource efficiency have been viewed as 'material considerations'; as important elements to be factored into decision-making for individual development proposals and in the preparation of spatial plans.

However, the two aspects of 'holism' and 'subsidiarity' embedded in the 'new' concept of sustainable development have proved to be a little more challenging for planners and other interests involved in development. The emphasis placed on a more holistic (or integrated) approach to analysis and decision-making has challenged planners to seek out 'coordinated' action to address the tri-partite environmental, economic and social concerns of sustainable development. It has also sat a little uncomfortably with planning's long-standing commitment to 'balance' (weighing costs and benefits and seeking 'compromised' rather than 'win–win' outcomes). Similarly, the call to 'think global and act local' has reactivated consideration of how the planning hierarchy (see Chapter 6) and spatial scales of policy and implementation need to work effectively and in concert, in order to deliver sustainable outcomes. Given the compression of time-space and the globalisation of economic and social life, efforts to develop a sustainable pattern of growth and change are made ever more difficult. Marrying these changes with sustainability's emphasis on more bottom-up decision-making has just added to the inherent tensions involved.

The Rise (and Rise?) of Sustainable Development

Over time, sustainability has become an explicit structuring concept in planning across the globe. However, the term and how it is defined or operationalised varies from place to place and over time. It is also mutable, reflecting social and cultural understandings as new information and ideas emerge. Given its radical implications and challenges for social and economic activity, how sustainability is defined and how it responds to changing interpretations becomes an intensely politicised process. In general terms, on the one hand, most politicians seek to insert the term into their rhetoric without a great deal of critical self-reflection. On the other hand, in planning practice, different interests manoeuvre around it in order to gain the initiative and to delimit sustainability in ways that will support their own arguments and intentions. Whatever way it is defined or used, there is no doubt that it has become one of the dominant discourses currently shaping planning debate. How has it arisen and from where did the concept come?

Although the concept of sustainable development has emerged over recent years, it grew from a long history of ecological (and other) thought. Robert Nisbet's (1973) review of Western social and political philosophy identified a number of threads to this 'ecological community', and others (e.g. Dobson, 2007; Pearce et al., 1996; Scott and Gough, 2003) have more recently emphasised the criss-crossing of different ecological, political, social and economic ideas. Indeed, many contemporary discussions of the term draw on the roots of 'ecological' perspectives in native/Eastern philosophies. Thus, for example, quotations by Chief Seattle of the Suquamish tribe are regularly sprinkled around web sites and articles on sustainable development. A similar 'ethical' dimension on the human–nature relationship can be found in religious teachings from across the globe. For example, in the Western Judeo-Christian tradition, many writers and practitioners have followed the way of St Francis of Assisi, who considered all living things as sacred and part of the family of God. For St Francis, the world of God and the world of nature were one; thereby exploding the dualism of nature–society and problematising the (unsustainable) exploitation of natural resources by humans.

Influences from political philosophy and practice can be seen in the anarchist or communitarian thought of Peter Kropotkin (1912), with his call for mutualism, self-sufficiency and cooperation. Similarly the long tradition of socialist and communist thought and praxis is largely built on conceptions of equality and provision for social need that characterize part of modern definitions of sustainability. Strong trace elements of this were present during the turbulent years of the English Civil War when the levellers and diggers sought to 'turn the world upside down' and initiate a system of common ownership and political and economic equality (Crouch and Parker, 2003). Moreover, interpretations might also be influenced by the underlying worldviews held by different groups of people. According to Scott and Gough (2003), this hinges on the degree of individualism or collectivism believed to underpin actions and the likelihood of our ability to effect change.

Although all these ideas are part of the network of relevant concepts, the first mention of 'sustainable development' as a distinct entity is correlated with another political force that gained momentum in the late 1960s and 1970s: the environmental movement. A string of academic and popular texts, such as Rachel Carson's *Silent Spring* (1962) and the Club of Rome's *Limits to Growth* (Meadows et al., 1972), were entwined with environmental agitation, protest and organisation. It was in this context that formal governmental institutions eventually acknowledged and made some attempt at addressing the major environmental trends and issues that had first ignited the protest movement. So when, in 1983, the United Nations General Assembly asked Gro Harlem Brundtland to chair the World Commission on Environment and Development (WCED), the ingredients were already in place. The subsequent report of 'The Brundtland Commission', *Our Common Future* (1987), merely distilled and 'legitimised' many of the ideas that had been circulating, and had been contested, for centuries and certainly in the preceding two decades.

The Brundtland Report is worth quoting here using a more extended definition, which states that sustainable development is:

> the ability of humanity to ensure that it meets the needs of the present without compromising the ability of future generations to meet their own needs ... Sustainable development is not a fixed state of harmony, but rather a process of change in which the exploitation of resources, the direction of investments, the orientation of technological development and institutional changes are made consistent with future as well as present needs. (WCED, 1987: 9, 43)

This places emphasis on meeting current and future needs (rather than 'market' demands) and sees sustainability as a process of change (of 'development'). However, it is unashamedly anthropocentric (i.e. human needs come first), as are almost all governmental or policy definitions that spring from Brundtland. This contrasts with more radical interpretations that are more earth- or eco-centred, such as James Lovelock's Gaia hypothesis (1979) or Arne Naess' *Deep Ecology* (1989). Hamm and Muttagi (1998) go part way between the weaker or stronger (or deeper green) views, arguing that sustainable development revolves around 'the capacity of human beings to continuously adapt to their non-human surroundings by means of social organization' (Hamm and Muttagi, 1998: 2). This places the emphasis on humans to behave and organise responsibly. Recent (re)constructions of the concept have tended to play out this tension between 'strong' and 'weak' versions of sustainability, and between more holistic and partial recognitions of the implications of environmental change and resource depletion. This is reflected in official UK government versions where the 'need' for economic *growth* was moderated towards a more balanced 'economic development' and then modified again towards a presumption in favour of sustainable development, with the emphasis on 'development'. Despite the changes in specific words in government reports, it is in the praxis of sustainability where the discourse is in a constant state of interpretation and reinterpretation. This is where planners are central to the ongoing meaning(s) of the term and to the particular outcomes that are negotiated and implemented.

More recently, a degree of consensus has emerged globally over the impacts of human activity on the global climate, with possibly severe consequences. As such the climate change agenda has become important and now rivals, as well as overlaps with, the primary aim of sustainable development. Indeed in December 2007, a supplement to the UK Government's *Planning Policy Statement 1* was added specifically on climate change, outlining the implications for planning. This guidance was revised in 2010 and demonstrates how rapidly the climate change discourse has provided a key justification for promoting sustainable development. It not only consolidates the call for sustainable forms of development but also places emphasis on the reduction of carbon dioxide (CO_2) emissions as the main issue:

> The Government believes that climate change is the greatest long-term challenge facing the world today. Addressing climate change is therefore the Government's principal concern for sustainable development. (Department of Communities and Local Government (DCLG), 2007a: 8)

As such, and beyond that already discussed, planning has a role to play in terms of both mitigation and adaptation to climate change and its possible effects. This involves shaping and locating development and other practices to ensure that decisions that are taken provide 'resilience' to change – defined as the ability of a system to resist, rebound from or absorb actions without fundamental change. The notion of 'adaptation' is where social, ecological, or economic variables are altered to ensure that society operates within sustainable limits. While 'mitigation' involves seeking ways to reduce or prevent CO_2 emissions and to increase the capacity of carbon 'sinks' which can absorb such emissions – some of these solutions are behavioural and some are technological. Many have spatial implications and demand robust thinking, over sustainable land use for example (see Foresight, 2010), and often infer regulation of some form.

Deconstructing Sustainability: How many Key Principles?

Despite the inherent process of social construction and reconstruction in debates over sustainable development, a number of key principles can be identified and outlined. These components are regularly included in debates over sustainability and move the concept through the application of related policies, techniques and practices.

Environmentalism

Environmentalism incorporates the idea that the full environmental costs and benefits should be considered in any decision-making process. It places 'stewardship' and protection of the physical environment at the heart of the sustainability debate and asks humans to move towards an eco-centric perspective in their relationship with the physical environment. Of course, there are many different shades of 'green' invoked in this movement from an anthropocentric to a more eco-centric view of the world, and there are many forces working against it. As mentioned earlier, planning has a long history of conservation, protection and resource management that has sought to maintain and 'husband' key environmental assets such as habitats, landscapes, water resources, open and agricultural land, air quality, historic buildings and the like. It has not always been successful in these objectives, but there is no doubt that environmentalism has played a significant role in shaping planning policies and practices. These influences stem from the conservation and preservationist movements of the late nineteenth and early twentieth century and the establishment of organisations such as, in the UK, the Campaign for the Preservation of Rural England (by the planner Patrick Abercrombie), the National Trust and latterly in the 1960s by the Friends of the Earth and Greenpeace.

Futurity

'Futurity' is a central idea behind Brundtland's definition of sustainable development. It demands that we consider the long-term consequences of any decision or action taken today in relation to the likely needs of future generations. Environmental quality and the future of the planet are obviously important to this idea, and it has also been constructed with an eye on the quality of life of future (human) generations. This coincides with the characteristic of much planning which takes a long-term view on future needs and the implications of decisions taken in the present. The very act of creating a plan is a future-orientated management device, seeking to reconcile the needs and demands of the current generation with the perceived future needs of those interests and the generations to come.

The futurity dimension of sustainable development also emphasises a precautionary approach to decision-making in which preventative or avoiding action is taken in advance of scientific certainty (in the case of existing environmental harm) or decisions are deferred until scientific proof is available (in the case of potential environmental risk). This concept has been a difficult one for planners (and others) to adapt to, as it operates in tension with the tendency for plans and planners to estimate and provide for future states. There are other means of precautionary delay (e.g. through phasing of development or subjecting development options to further scrutiny) or of restraint (e.g. against development on floodplains or in areas of poor air quality). However, these considerations and their spatial implications have not had an easy passage when development pressure is high or when the long-term impacts are not well understood, perhaps particularly by politicians acting as planning decision-makers.

Development

The two words 'sustainable' and 'development' (see also Chapter 19) are put together to stress a process of change and improvement. It suggests that 'staying still' in a cocooned world of environmental protection is not an option. It introduces a more anthropocentric spin on 'environmentalism', implying that there is a requirement to undertake further economic and physical development to provide for future needs. In the same way that 'environmentalism' has many shades of green, so too does 'development'. In this light, the UK government has often reconstructed it at one extreme, in terms of maintaining 'high and stable levels of economic growth and employment'. Other lobby groups and organisations have tried to promote an alternative conception of development which is based on a more holistic and cyclic view of resource flows in the built and natural environment. More recently ideas such as 'ecosystems services' have re-emphasised the connectedness and interdependencies of different actions and functions. Development forms like 'low impact' or 'zero-carbon' housing or 'eco-industrial' estates have been proposed and operationalised as indicative examples of what can be done to 'live lightly upon the earth'. Such ideas and examples have tended to remain marginalised by a capitalist development industry that has been unable (or unwilling) to adapt to such a regime. While it could be said that they remain something of a sideshow, these ideas continue to challenge the technocratic discourses of sustainability that currently tend to dominate planning policy debates. The deeper green and more integrated views are also becoming more influential as governments grapple with the implications of the climate change agenda.

Equity

The 1992 Rio de Janeiro Earth Summit played a key role in reconstructing the concept of sustainability. A major non-governmental organisation (NGO) presence at the summit (and during the preparations for it) managed to lend this principle more weight than it had hitherto achieved in sustainability discourses. The notion of equity in this context is based on the view that inequality in the distribution of the earth's resources and wealth is a fundamental product and cause of unsustainable development, and that, in order to address this, decision-makers at all levels should be required to consider the distributional consequences of their proposals and actions. Moreover, when considering actions across space, a responsibility towards other less-developed countries or areas should be factored into decisions.

By raising these types of issues, the sustainable development debate opens up clear linkages with a long history of social-reformist ideas and political/social movements mentioned above. How this principle is operationalised in both general and specific cases will also reflect the continuum of thought that pervades the political spectrum of the people and states involved and the way that they interact. Planning practice opened up to some of these ideas from the 1970s in the UK, when the inequalities in planning and urban development processes were exposed to significant critique. However, the 'equalities issue' has tended to be institutionalised into relatively anodised and generalised 'equality considerations' that get attached, perhaps as afterthoughts, to planning appraisals and planning application decision reports. The 1998 Human Rights Act is one example of this (see Parker, 2001) where a potentially powerful vehicle for maintaining individual rights has been absorbed almost unnoticed into day-to-day planning practice.

Participation

One of the main outcomes of the Rio summit was the *Agenda 21* statement. This document established the meaningful participation of individuals and groups in decision-making and implementation as a touchstone of sustainability. Meaningful participation and involvement of citizens and 'stakeholder' groups is based on a number of rationales, including the need to:

- tackle environmental problems at all relevant levels (including the local);
- build consensus between all key interests (including those normally marginalised);
- spread ownership of sustainable development down to individual communities;
- allow local solutions and decision-making, where appropriate (i.e. political and economic subsidiarity).

This view of participation is not without criticism and needs to be treated with caution given the clear inequalities in power and influence held by different stakeholder interests (see Chapter 9). However, it does emphasise the importance of this dimension to many people working with(in) the concept and practice of sustainable development. Concern with public participation in planning decision-making predated the Earth Summit by some years in UK planning practice (see Skeffington, 1969), and ever since the 1960s planners have been grappling with ways to engage more directly and effectively with a range of community interests. This process has met with its fair share of failure, frustration and tokenism, but progress has been made and lessons have been learnt (see, for example, Brownill and Parker, 2010; Healey, 2005; Rydin, 2010b). The participatory reforms to the English planning system in 2004 (Her Majesty's Stationery Office (HMSO), 2004) were predicated on a need to ensure speed and transparency of decisions and yet also to reaffirm sustainable development as an overriding priority. Even though such 'reforms' may seek to promote aspects of the 'ideal' of sustainability, other priorities and their interpretation on the ground – such as competitiveness arguments (Chapter 17) – have often tended to win over environmental ones.

This list does not exhaust the components that go to make up the discourse and praxis of sustainability. As mentioned above, the essential requirement for holistic/integrated thinking and action is a challenge that planners are struggling with at present. This is somewhat ironic given the reputation of planners as reticulists and given recent initiatives towards developing an integrated 'spatial' planning approach. Similarly, the idea of 'environmental' (or 'natural') capital (see Chapter 16) has been tried and, to some extent, found wanting as a 'sustainability tool' to manage the release of land for development. The nebulous but very potent concept of 'quality of life' is also regularly mentioned in policy-based definitions of 'sustainable communities', while the precautionary principle and ecological diversity or biodiversity are also terms that regularly get inserted into debates about sustainable development. As suggested earlier, these principles, concepts or tools are deployed by a range of interests to construct and operationalise particular discourses of sustainability. The above 'conceptual map' merely illustrates the inherent malleability of the term, something that is illustrated well when we look at examples from planning practice. This malleability is both a strength and a crucial weakness when placed into the tough and politicised world where development and economic growth are pushed hard by an array of business and other interests.

Sustainability in Planning Practice

Increasingly sustainability has been a prime justificatory tenet of the planning system and this is reflected rhetorically in many of the major policy documents present in the UK and elsewhere. In the English planning system for example, the overarching purpose for planning was set out in *Planning Policy Statement 1*, which stated that: 'Sustainable development is the core principle underpinning planning. At the heart of sustainable development is the simple idea of ensuring a better quality of life for everyone, now and for future generations' (DCLG, 2005, para. 3). This reflects the imposition of a specific but broad use of the term sustainability, which simultaneously marginalises other 'deeper' definitions and allows a great deal of scope for interpretation and application in practice. This frame setting by central government policy limits how far planning processes can construct other interpretations of sustainability, but it does not fully determine this.

There are other influences structuring the process including the planning system's prevalence to see sustainability in terms of 'balance'. Balancing of different impacts and interests results in the implementation of a form of pragmatic or 'net sustainability'. This tends to downplay the tensions within existing dominant development processes, and involves a trading-off of sustainability considerations against other 'material considerations'. In this way the holistic/integrated/reticulist potential of spatial planning is denuded as historic processes and procedures reimpose themselves in day-to-day practice.

Murdoch (2004) discusses how discourses of sustainability have been reflected in particular technologies and practices in planning, in relation to the rise of the policy for brownfield housing development in the UK. What is significant about the priority placed on brownfield redevelopment in the mid 1990s is that it quickly became a surrogate concept for sustainable forms of urban development – rather than being seen as one part of an approach to efficient and sustainable land and resource use. Sensitive to the influence of a well-organised and proactive countryside conservation lobby (fronted by the Campaign for the Protection of Rural England (CPRE)) the incoming Labour government helped construct a policy framework that was summarised by the then Housing Minister, Nick Raynsford, as 'brownfield first; greenfield second' (2000: 262). As Murdoch suggests:

> The new political rationality that came to dominate the planning for housing arena in the late 1990s effectively involved the rather selective appropriation of elements within the sustainability discourse by CPRE and central government. Sustainability was now interpreted not in its usual sense as a balancing of economic, social and environmental criteria with the development process but rather as the re-development of already-developed land. (2004: 53)

This narrowing-down or selective use of the label of sustainability was clearly evident in the 2000 version of the Government's *Planning Policy Statement 3* on housing (Department of the Environment, Transport and the Regions (DETR), 2000) when it offered as guidance: 'to promote more sustainable patterns of development and make better use of previously-developed land', in which, 'the focus for additional housing should be existing towns and cities'(para. 1). The emphasis on re-using brownfield sites has consistently been used since 1997 as the touchstone of sustainability in the UK. The main policy statement about housing land-release was clear that the government saw *sustainable patterns of development* as (DETR, 2000, para. 21):

- concentrating most additional housing development within urban areas;
- making more efficient use of land by maximising the reuse of previously developed land and the conversion and reuse of existing buildings;
- assessing the capacity of urban areas to accommodate more housing;
- adopting a sequential approach to the allocation of land for housing development;
- managing the release of housing land;
- reviewing existing allocations of housing land in plans, and planning permissions when they come up for renewal.

Although the efficient use of land dominated this guidance, other sustainability concerns such as the greening of the development process, public transport provision and the conservation of heritage and ecological resources were included in *Planning Policy Statement 3*. However, they did not receive the same kind of emphasis and were treated as subsidiary additions to the main policy criteria of reusing previously developed land. This is also reflected in the way in which the policy for housing development was to be taken forward into implementation or, as Murdoch (2004) suggests, 'materialised' through 'governmental technologies'.

The 'urban capacity study' was the main planning technique (or 'technology') advocated in this guidance in order to secure the(se) aims of sustainable development. Arising from earlier work on *environmental* capacity studies, urban capacity studies were developed by an emerging network of environmental consultants, environmental pressure groups (including Friends of the Earth and the CPRE) and some proactive local planning authorities. This required local authorities to identify and quantify previously developed sites in their (urban) areas that could accommodate new housing development. As Murdoch shows in his study, local authorities were able to bend the technique in various ways to bring the findings more in line with local political priorities. In terms of theory, it illustrated that:

> The operation of the capacity study is part of a network-building process, one that ties the deliberations of central policymakers to the many urban sites that are to be enrolled into the planning-for-housing process. In effect, the capacity study allows the government network to place a (particular kind of) 'sustainability frame' around local development decisions.

But also that:

> By selectively appropriating elements of the guide, local planning decision makers can steer their conduct in line with local, rather than national, sensibilities. (Murdoch, 2004: 55)

This shows how attempts to work towards sustainable development forms can be frustrated or diluted if the governance and policy environment is not appropriate and where understandings of sustainable development are inadequate. Urban capacity studies are not the only technology of governance constructed to take forward the discourses of sustainability. Other examples, with their own actor-networks (see Chapter 4), include: environmental impact assessment; sustainability appraisal; the listing of historic buildings; the designation of ecological or heritage areas or sites; environmental audits; the application of sustainability indicators; the Building Research Establishment Environmental Appraisal Method (BREEAM); carbon reduction strategies; and lifecycle analysis. Indeed there is some indication that some of these technologies are taking over from urban capacity studies in more recent UK government legislation and guidance (for example, the requirement for new learning development to be built according to specified sustainability standards).

Conclusion

Our account of sustainability and planning demonstrates the multifaceted nature of the idea and of the difficulties in implementing deeper or stronger versions of sustainable development. The discourse and reflection of sustainability in policy shows it is in flux; it is in a constant process of construction, stabilisation and (re)construction, aided and abetted by various technologies of governance and as new evidence and political consensus is brokered. It is also used in conjunction with other ideas and concepts – some of which are discussed in this book.

Irrespective of this malleability, the concept has some resonance that shapes planning policy and practice. At its core sustainability is about ensuring that decisions taken today are justifiable and beneficial in the long term and that short-term motives, such as profit making or responding to a need or want of today, cannot be the overriding factors in our decisions. The calculations and decisions made need to be based on good environmental evidence and understanding, even if some of these are precautionary and experimental. In general terms planning needs to become more integrated and efficient in thinking about resources and the social impacts, the economic priorities and the environmental goods involved. In these ways, the efficient use of land, water and energy, the promotion of greener urban forms and shifting peoples' attitudes are necessarily reasserted as prime considerations for planners.

FURTHER READING

The topic of sustainability has understandably spawned legions of papers, reports, books and other articles – many of which focus on a particular policy area, or technique or aspect of sustainability. However, the Club of Rome's *Limits to Growth*, published in 1972, captures a moment in time when the imperative to check and rethink the resource use of humans echoed loudly around the world, and a 'thirty years on' report was also produced in 2002 (Meadows et al., 2004). For the UK, the policy document *Planning Policy Statement 1* and the climate change supplement make interesting reading (DCLG, 2005, 2007a), before their incorporation into the coalition government's *National Planning Policy Framework* (DCLG, 2011). Scott and Gough (2003) and Hamm and Muttagi (1998) set out the need to educate and learn about sustainability and cities; Meadowcroft (2000) highlights how sustainability has become a dominant justification for planning activity; while Rydin (2010b) critically evaluates different aspects of sustainability planning. Murdoch's (2004) paper gives us an example of how sustainable development efforts can be manipulated in practice, and the Owens and Cowell (2002) text adds to this.

4 NETWORKS

Related terms: resources; cross-boundary ties; policy formulation; complexity; systems; actors; hybrids; flows; mobilities; relations

Introduction

Planners play an important role in structuring and shaping policy and effecting decisions through networks of relations and resources. The term is typically applied to a physical arrangement of elements as an interrelated 'grouping'. However, the common usage of the word network does not fully convey its range of meaning and importance and this is insufficient for planners and others engaging with development and spatial policy. Arising from this, questions need to be posed about what constitutes networks, how they operate and the relevance of this concept in relation to planning practice.

The analysis and consideration of the shaping of networks as 'sets of relations' is increasingly seen as an important part of planning theory and practice. This chapter outlines the relevance of networks to planning practice, as well as setting out some examples of how network perspectives have been applied to planning. An understanding of networks is not straightforward, however; not least because there are different 'types' and conceptualisations that are of potential relevance. Each has its own import and internal complexities. In this chapter, we consider definitions, types or purposes of networks, their components, and how different conceptualisations of networks and their analysis may be useful for planners.

Definitions of Networks

How relations and decisions are negotiated by public bodies, local groups, developers and others can provide the material and constituents of a network. The role of various networks in shaping planning outcomes is discussed, for example, is by Bijker and Law (1992), Murdoch (2006) and Evans et al. (1999), and such assessments provide much of our focus in this chapter. Beyond obviously apparent (or physical) networks, such as transport networks or communications infrastructure networks, we are also interested in social and cultural dimensions that act to constitute planning and development processes. That is to say, how combinations of social and technical entities are assembled through network processes and act to 'produce' planning outcomes.

There are numerous definitions that relate to networks, reflecting different paradigms through which the networks metaphor may be viewed. Some forms of network analysis can challenge presupposed understandings by including in the frame of analysis relations and interactions that break away from both the present day and the narrowly spatial or Cartesian accounts of the visibly observable. As a result, awareness of a range of influences and information that stretch across space and time can be extended. Before developing this, the initial or surface scrutiny of networks recognises them as 'groups bound by some form of connection or relationship', or alternatively as 'groups which share some similar function or design needed to perform a particular task'. While it is clear that planners work with others, a rather deeper assessment of assemblages of materials and people involved in shaping the environment involves looking at the relations of power and influence that the different actors hold. In this, we think, a richer conceptual view that extends those simple definitions of networks comprises actors who instigate or receive exchanges as well as the intermediaries who 'carry' those exchanges (and sometimes affect or define the interactions between actors). The 'network' is therefore either defined as a set of relationally tied actors, or a wider and changing assemblage of people and other beings or things (sometimes referred to as 'actants'). Our interest then is not only restricted to *who* is involved as part of a 'planning network', but also *how* such networks operate and *why*. We begin to discuss these points with reference first to *social networks*.

Social Networks

Social network ideas provide a frame for the analysis of human organisation. The relevance of social networks will vary and they may or may not affect attitudes to planning issues or policy formulation. It rather depends on the context, and moreover, how networks are viewed and understood by planners (and others). Relations between actors implicated in a network are seen to be what constitutes the network and gives it force. Hillier sees (social) networks as, 'relational links through which people can obtain access to material resources, knowledge and power' (2002: 113). Attributes of individuals are viewed as less important than their relationships and ties with other actors within the network, although such characteristics may influence relations. Social network theory tends to be largely aspatial, deriving from its roots in sociology, and can miss how such networks may extend across space and act to link 'distant' people and places. This spatial dimension has important ramifications for place and people relations and is significant when considering patterns of development and human behaviour.

The focus on networks downplays individual agency to an extent, implying instead that the aggregated structure of the network constructs its trajectory. This view is not always plausible, and in rebalancing the role of the individual, Healey (2005) outlines why social relations may be important in forming personal identity and how such networks, as 'grouped agency' or 'centred subjectivities', shape places and the way that spaces are used and perceived. Parker and Wragg (1999) highlight how individual actors can destabilise and redirect networks, highlighting the potential of 'networked agency' and the role of network builders in steering the process. Hillier (2000) points out how networks can be used to subvert planning processes. In the same vein, meaningful design and operation of public involvement in planning requires an understanding of the types of group that exist and how they may be impacted on by policies.

Social network researchers have tended to focus on the observable relations and outcomes that a group of people act to co-design. This is a useful way to map out stakeholders in planning and for planners to facilitate or advise such networks. However, there are other benefits to planners of developing a network understanding, to which we now turn.

Chapter 4

Planning, Networks and Actor Network Theory

The post-structural turn reopened theoretical debates that challenge the simplification of binary divisions and dualisms, and has re-emphasised technological or material influences on society. Assumptions about the nature–society divide and the clear distinctions typically made between humans and non-humans were deconstructed and found wanting. It was not long before these post-structural insights were being applied to the social sciences and, by association, to the theory and practice of spatial planning (Hillier, 2007; Murdoch, 1997b, 2004). As such, social network theory has been challenged as an incomplete perspective, with post-structural thinking inserting discussions of hybridity, fluidity and contingency and allowing for the role and effect of technical and other non-human 'actants' on supposedly 'social' networks. Networks tend now to be regarded in terms of a wider 'assemblage' of the human, non-human, material and technological that is orchestrated to particular ends.

Reflecting this shift in thinking, actor network theory (ANT) has become an influential approach to the study of space and networks that shape space: 'ANT is advocating ... a socio-philosophical approach in which human and non-human, social and technical factors are brought together in the same analytical view' (Law, 1999: 1). This approach extends the scope of network theory, challenging traditional conceptions of space used by many planners, in which distance equates with space and where people, places and things can be located by reference to abstract coordinates and mapped onto a simple cartographical representation. Networks from the ANT perspective can be conceptualised as stretching across space. The shape or extent only being limited by the influence of actors who can 'act at a distance' and vicariously impact on places and other actors/actants through the 'reach' of the network and the relations circulating in the network. Individual actors operate as brokers by bridging within and across different networks. This filling of 'structural holes' (Burt, 1992, 2004) is a potential role for planners: acting to negotiate and bring together different interests (or networks) and to broker new network relations that promote planning goals such as sustainable development. Moreover, the identification of relevant actors and intermediaries might be widened to include natural and other 'actants' beyond the human. This conception underpins Cannon's

(2005) 'adaptive management' approach to brownfield remediation, in which planners are urged to 'think like a contaminated site'. It also supports Armitage et al.'s (2007) arguments for 'adaptive co-management' where institutions, policy and knowledge are tested and revised in an ongoing way by the range of actors affected.

Researchers have recognised how networks can act to crumple or 'pleat and fold' space and time. By taking this effect of networks into consideration, planners can think about policies and plans for particular bounded areas – asking, for example, how discrete areas are connected and influenced by each other. This kind of analysis has been made possible primarily because of an increased awareness of the complex assemblage of relations that structure and shape places, and the socio-economic and environmental attributes affecting quality of life and sustainability, as key objectives of spatial planning. In other words, thinking about a given planned area should involve looking beyond the bounded space defined by administrative borders to include the relations, flows and effects from both within and beyond that district or region. For example: How is the economy of the Thames Valley affected by decisions made in Beijing, Brussels or Baghdad? How is economic activity in one district affecting groups who live in another? How might national or international airport policy impact on a whole range of actors, actions and possibilities? Taking this widened conceptualisation of actants further also means actively considering the 'actions' and impacts of, for example, rivers, wildlife or technology in spatial planning processes.

Thus serious questions are raised about how functional cross-boundary connections are maintained or recognised in planning. How do we need to think about networks in order to plan more effectively? This leads to the ideas covered in Chapters 5 and 11 and the difficult but important perspective that plans and planners actually operate in 'open systems', despite historical and ongoing efforts to close off or narrow the scope, relevance or legitimacy of wider considerations and claims made to planners and politicians.

Networks are seen here then as a complex or heterogeneous assembly of relations/resources. This implies that planners should recognise the way networks or associations are in flux and extend across boundaries, reaching wherever connections or ties are present. The idea of the 'glocal' reflects the sense that actors and resources are circulating simultaneously at different scales but may be bound by some common interest or need. In order to facilitate interactions and relations, networks need intermediaries. These are resources used to facilitate network interactions, for example texts, machines or money (see Callon, 1991). In many instances planners may themselves have different roles and become intermediaries in networks, actors or 'network builders' (i.e. when developing a planning policy).

In this view, the network is the extent of relations aligned or organised to provide a particular outcome. For example, the outcome could be to produce car engines, or to develop an implementation strategy, build a bridge, manage a national park or to protect an endangered species. However, besides identifying the importance of relations and ties, there are several other aspects of networks that need to be highlighted. One of the most important for planners to consider is the *reach* of the network: that is, the origins and extent of linkages which impact on economic, social and environmental relations within, across and between places.

Acknowledgement of relations and heterogeneous connections between people and other aspects of place, such as buildings, artefacts and less tangible components such as memories (see Dean, 1996), intimates a link to the development and maintenance of social and human capital. We can begin to see, therefore, that the 'network' may be more than the sum of its parts – that relations and exchanges among actors generate a power and 'direction of travel'. This may also have spatial and other policy connotations and furthermore, such connections may imply a drawing together or indeed a fragmenting of spaces and places, which, in turn, requires new strategic and cross-boundary thinking, coordination and understanding. It represents nothing less than a fundamental rethinking of the dominant planning paradigm.

When looking at the context or role of the network we should also consider what the *purpose* of the network actually is and how such networks operate in terms of planning issues and policy formulation. Therefore we should consider network direction and trajectory and where these might militate against preferred 'planning' processes or outcomes. In the light of this, we now turn to consider how networks operate, and why, starting with the distinction between policy and issue networks and how planners might be involved in these.

Policy and Issue Networks

An awareness of the characteristics of networks that are formed around policy areas and issues is highly relevant. Many planners will work with set processes and procedures and follow programmes and schedules handed down by authoritative and legislative sources. Conversely they may be involved in unforeseen or otherwise sensitive deliberations about policies or decisions where exercising discretion and negotiation are important. In such roles, planners are implicated in both *policy* networks and *issue* networks (Marsh and Smith, 2000). Murdoch (1998) points out that issue networks are likely to be negotiative (i.e. open and fluid) while policy networks will tend to be more closed or 'prescriptive' – with less opportunity to affect process and outcome. *Policy networks* may have developed accepted 'rules' which clearly demarcate legitimate behaviours and relational order: in such circumstances the network may be guarded via 'obligatory passage points', which ensure that particular processes are followed and that prescriptions are 'shaped' appropriately. Such conditions ensure control and conformity (e.g. types of information being disallowed, deadlines being imposed, the format of inputs being set). Many planning activities such as the formulation and scrutiny of development plans have been managed through such policy networks.

This situation comes close to the idea of a 'policy community' where a set group follows a more or less strict process and particular types of resource, information and actor are considered legitimate (Jordan, 1990). However, it is a notion that becomes problematic when ANT is used to analyse such contexts. Rarely are such communities consistent or able to fully control the process of policy formulation, providing a warning to politicians and planners who seek to narrow or 'close off' policy processes or decision-making too tightly. Moreover, to do this threatens the legitimacy and 'implementability' of such policy, and the network may be more easily 'destabilised' (Parker and Wragg, 1999).

Issue networks are typically more *ad hoc* or loose networks that form around a particular topic. Planners may be involved as part of an emerging issue, or the activity may be an innovative part of a local authority approach to planning. As that issue is identified and a solution is designed or calculated, actors come together and seek to determine an outcome. The information, power and other abilities of actors are crucial here. In addition, the agency and resources that are brought to bear, or which are recognised or omitted, can markedly influence the trajectory of the network (see Davies, 2002). Possibilities for different actors and outcomes, knowledges and methods are wider and even less predictable or 'controllable' in issue networks. Issue networks tend also to be more reactive associations that respond to a situation or proposal. In contrast policy networks tend to be longer lived or act as 'standing committees' that oversee more established processes of decision-making, often with some 'official' or otherwise recognised status. Both types coalesce to provide a particular (non-determined or unpredictable) outcome that involves negotiation and the deployment of resources to achieve success.

An alternative theorisation of this process is contained in the ANT-inspired translation theory account. This account explains how networks tend to pass through idealised 'stages' to stabilise and reach agreement. The stages are: 'problematisation' of the issue; development of the necessary network (the 'interessement' stage); 'enrolment' of necessary actors and resources; and, lastly, 'mobilisation' of the network to implement the preferred solution or policy (see Callon, 1991; Murdoch, 2005). However, networks are never fully closed, nor are their outcomes fully predictable. It may be difficult to fully comprehend the extent of the relations involved too. Yet, typically in policy networks, efforts will be made to exclude some actors or to set up conditions of entry or other barriers – such as professional language. Otherwise, a good deal of behind-the-scenes activity will take place to organise and 'simplify' outcomes, which often happens during policy formulation or when deliberating over large development proposals. One of the public frustrations concerning planning has been the way that the 'policy community' process typically arrives at decisions. However, attempts to close down options can result in subsequent contestation, possibly when an unforeseen event or new information comes to light.

Planning and the Challenge of Networks

It may be difficult for planners to reconcile the implications of network understandings with practical demands on their time, energy and other resources. Nevertheless, it is important to understand the environment in which 'planning' is acted out and the possibilities and complexities of that environment. As such we have adopted here a post-structuralist and heterogeneous view of networks that can be defined as 'hybrid assemblages of relations and materials of people, resources and interactions that are stretched across space and time'. In other words, networks are fluid, requiring maintenance, and difficult to fully comprehend or control. While planners may be interested in physical networks of things such as roads and rail connections, or even groups of people *qua* networks, there are other linkages and interactions that act to shape and reshape planning and spatial outcomes. Criticisms of planning and implementation failures in the past have arisen partly because of poor understanding of the interrelationships and contingencies existing between and among groups and places. So planners need to become active negotiators and network operators, as well as the more obvious technical or process 'experts'.

In adopting a network perspective in planning key questions begin to arise. Who put the plan together and how? Who has 'agreed' to it? What interests have had influence over the authors and why? This then leads to questions over how 'agreed' points will actually go forward and be realised. A factor that arises here is the common lack of understanding and consensus actually achieved in devising a plan or a strategy, despite claims which are often made to the contrary. In our terms, the network building may have been instrumentally efficient (for someone), and passed through a series of required processes and stages or passage points, but: in whose interest and at what cost (to others)? Who has actually engaged and understood the implications of the plan? Will it be sufficiently supported to be implemented? By adopting this deliberately critical and questioning stance we hope to invite reflection around how plans 'fit' the complex of people/place relations that they seek to affect.

Our contention is that plans and policies may not actually be achievable as a result of dissonance or disagreement. Affected parties may be unwilling or unable to adopt or accept the implications of the strategy (and hold some power over its implementation). Furthermore planners are often unable to cope with rapidly and iteratively changing circumstances and the extensive possibilities that ongoing interactions and negotiations imply for their 'patch'. This is one of the factors that accounts for a net failure to fully engage with interests and other actants, and to anticipate needs, outcomes or otherwise prevent policy failure.

Beyond the processual aspect of network building lie the issues of complexity and change that have yet to be fully reconciled with planning practice (cf. de Roo and Silva, 2010). Many plans and policies have been designed as rather inward-looking strategies, or else they are intended to shape a particular aspect of the social–environment–economy nexus (e.g. promoting development or the protection of open space). In doing so, such plans rarely reflect the complex relations and conflicting forces that shape places or affect policies devised in planning offices.

As early as the 1960s, Webber was challenging planners to confront this issue, arguing that:

> Possibly the meanest part of the task [of planning] will be to disabuse ourselves of some deep-seated doctrine that seeks order in simple mappable patterns, when it is really hiding in extremely complex social organisation instead. (1963: 54)

Area-based place planning has been criticised widely and a better appreciation of networks could help planners tap into, and play an active role in understanding and shaping relevant flows of knowledge, people and material. This is perhaps even more important in a globalised era of rapid technological innovations, advanced communications, trans-border investment and other dimensions that have served to alter time/space relations (see Chapter 11). In such a context, the role of effective planning has become more reticular as networks shape and influence places, developments, local economies and social dynamics (Castells, 1996; Graham and Healey, 1999). Societies and networks have globalised and speeded up. Given the unpredictable, and difficult to control, flows of resources, capital, people and information, it has become increasingly clear that planners need to better understand the way that networks operate, change and who/what are involved in these networks.

Application of Network Understandings

Our focus in this alternative conceptualisation of networks is how they can be of use to planners and aid in tracing and analysing the political dimension of planning practice. Such descriptions of network types reflect a necessary skill in identifying network relations and reach.

Given this context, one plausible role for planners is to 'map' networks in order to understand the way in which processes of planning and development take place, how they can be facilitated (in particular ways) and with what associated implications. This form of analysis might assess what resources and actors/actants are involved, or are likely to affect planning and development policy and decisions. This positions planners as network analysts and network builders, working to steer networks to agreed or otherwise 'sustainable' outcomes. It also encourages a widened perspective for planners, intimating that previously underplayed factors may be usefully brought into analytical view (such as the socio-technical, the agency of non-human actants and cross-boundary flows). However, there are serious practical barriers to such a shift, including extant planning cultures, the political and adversarial environment of planning practice, the relative introspection of local authorities and possible resource implications of such a shift. All of these provide significant inertias and excuses not to change or to instead pay lip service to a new network perspective. Given the above, it is acknowledged that many planning authorities now recognise the need for an understanding and engagement with existing policy and social networks. One example where planners in England have been explicitly directed towards engaging with existing 'community' networks is where public involvement is sought. The following quote on the development of community strategies in England illustrates the way that planners may be seen as orchestrators and analysts of networks in this instance:

> The involvement of local people is central to the effective development and implementation of community strategies, and key to change in the longer term. There is an often untapped pool of ideas, knowledge, skills, experience, energy and enthusiasm among individuals, groups and communities as a whole which, if realised, can be a real driver for change. Community strategies offer a fresh opportunity to put local people at the heart of partnership working and should be grounded in the views and expectations of those people. (DCLG, 2007b: para. 50; see also Doak and Parker, 2005).

This emphasises the resources that may be available and the need to draw these in and account for them in terms of steps taken, policies formulated or schemes that are designed. Local resources can be drawn into wider networks and assist in more robust and consensual proposals for change. In this process it is useful to think about any gaps, omissions or 'structural holes' that may exist in a network and to reflect on some important questions for planners to consider. For example, what role do planners play in situations where there are power differentials, or where known interests are not enrolled into a network? Furthermore, what relations and constraints exist between different scales of the hierarchy? (see also Chapter 6). And what other entities or actants might be implicated in such networks? Are they being assumed to be compliant? Are they predictable and able to be enrolled into the network? What are the consequences of leaving structural holes, or network gaps unfilled?

Thus, the interest in thinking about planners as networkers and the ontological extension of networks as heterogeneous assemblages, perhaps better and more convincingly corresponds to a more fluid and multipartite world, where different relations, artefacts and entities act to shape the conditions in which planners operate. Regardless of a widened conceptualisation of networks it is clear to us that some appreciation of the role of different actors and resources that shape planning processes and outcomes is integral to the work of planners. Awareness of the implications of actors, intermediaries and relations that shape the world and the environment is a necessary skill for planning effectively.

FURTHER READING

The general literature on networks and policy process is wide and it is useful to read Marsh's work (e.g. Marsh, 1998; Marsh and Rhodes, 1992; Marsh and Smith, 2000). A clear and useful account of the wider policy process is provided in Ham and Hill (1993). In terms of actor networks, Murdoch's (1997, 1998, 2006) along with Selman's (1999, 2000, 2001) work provides a good general account of the use of network analysis in planning, as does Hillier (1999, 2002, 2007). Parker and Wragg's (1999) paper about the potential destabilisation of networks is useful, and Tait (2002) and Tait and Jenson (2007) have applied a form of actor network analysis to local authority plan-making processes, showing tensions between central and local government. A more general collection of resources on ANT is provided by academics at Lancaster University (http://www.lancs.ac.uk/fass/centres/css/ant/antres.htm). Chapter 11 on mobility and Chapter 5 on complexity are also useful counterparts to this chapter as they highlight the need for planning practice to embrace contingency, change and global-local flows.

5 SYSTEMS AND COMPLEXITY

Related ideas: networks; relationships; emergence; learning; adaption; collaboration; co-management; co-production

Introduction

Jane Jacobs was someone who appreciated the complexity of the city. She viewed large scale settlements as being made up of many micro-level interactions.

> Cities happen to be problems in organised complexity ... they present situations in which half a dozen or even several dozen quantities are all varying simultaneously and in subtly *interconnected* ways. (Jacobs, 1961: 433, our emphasis).

Jacobs was influenced by her reading of mathematics and natural sciences debates in the 1960s, which were exploring complexity and increasingly using systemic modes of analysis. This infusion of scientific thought provided a new perspective through which to explore the dynamics of cities. The key word that Jacobs uses to express how the city appeared to work is 'interconnected'. This strikes at the heart of a systems view of the world; and if we add the word 'non-linear' to it we get the essence of complexity theory, that is, that there are no simple cause–effect relationships in the city; the relationships and 'causes' are multiple and moreover outcomes are often unpredictable. Early forms of planning with their associated top-down and often reductionist or essentialist solutions to planning challenges are thus problematised.

Viewing societies, regions or cities as 'systems', places emphasis on an integrated set of relationships and dependencies. Indeed many authors have stressed the importance of relations as central to the conceptualisation of complex systems. In this chapter we review some of the key ideas and approaches that have been driven by a 'systemic' understanding of the world and how planners have attempted to intervene through the use of systems thinking. This systems view was developed in the 1950s as part of a 'rational' approach to planning which required planners to try to understand places in a holistic way (see Chapter 2). However, such efforts were confronted by a number of practical and conceptual difficulties. For instance, the boundary question of 'what to include?' in the analysis (see Chapter 4), related questions of priority and magnitude of relations, and how to cope with the sheer number of variables and contingencies found in cities and regions. Such scale and measurement issues are compounded by resource and time-related difficulties; planning organisations or clients often do not want to pay for the holistic analyses implied by this; particularly given the apparent speed of change and mutability of people–place interactions in a mobile and hi-tech world (Urry, 2000b). In essence the world moves on while analysts prepare plans, often rendering the initial analysis and the resulting plan partial, incomplete and obsolete and therefore open to charges of wasting resources.

Early systems theory applications in urban and regional planning (see Chadwick, 1978; McLoughlin, 1969) set in motion a change in planning approach from one largely concerned with the built environment and design considerations, towards a planning that sought to shape and mould society more generally. This ambition was partly enabled by developments in science and technology, in particular, the rapid advances in computer power. It was further supported by the rise of planning as an established and state-endorsed profession and the consequently widened use of city and regional planning as a tool of public policy. Indeed the production of the influential *Limits to Growth* (Meadows et al., 1972) was an early example of an analysis that was based on environmental forecasting using large-scale computer modelling. Aspects of this approach are still to be found in modern planning around the world and for these reasons we include complexity, along with systems, as a key concept here. The inclusion of these ideas also questions simple solutions or interventions in one area or topic and prompts wider thinking about the interrelated repercussions and relations touched by specific actions, or decisions relating to land use and spatial outcomes. In this sense, the chapter is also closely related to our discussion of networks in Chapter 4.

Systems and Complexity

This approach to planning has generated significant debate over the practical and conceptual difficulties of adequately representing the world. The question of how to confidently model systems and to understand the effectiveness or impacts of decisions informed by systems theory is brought into view here. While there are numerous ways of thinking about both systems and complexity, the connections and relationship between systems theory and complexity theory need some discussion. Here, we outline the basic elements and assumptions made by systems theory before setting out the differences, shared aspects and extensions that the concept of complexity and associated theory has provided since the 1990s.

Systems theory

Systems theory in planning has been closely associated with modelling and the forecasting of change. Systems theory has historical roots in the sciences (particularly mathematics and cybernetics) and in 'general systems theory'. Systems can be said to exhibit a number of key features: they are said to have boundaries, defined functions, structural relationships and, crucially, that they are dynamic, that is, they change. General systems theory also claims that each variable in any system interacts with the other variables such that cause and effect cannot be easily separated. This point raises serious questions over deterministic forms of intervention, where decisions about particular outcomes may be rather simplistically linked to a 'probable cause'.

As with Jane Jacobs' view, the systems perspective regards 'the system' as a set of things that are interconnected. Therefore the interaction of those elements and moves to intervene in those structuring relations are seen as key for planners to understand:

> What makes a system is not just a set of distinct parts, but the fact that the parts are interconnected, and so interdependent. The structure of a system is therefore determined by the structure of its parts and their relationships. (Taylor, 1998: 61)

The implication is that such a perspective will help reveal levers and tools for strategic and local planning. Such analysis may also illuminate exclusionary effects of system operation in social or political terms, or the opportunity to insert normative aims that can be used to point or shape the system in a particular direction. The quote from Taylor also highlights a need to explain a little more about what a system actually is.

Chadwick (1978) was one of the leading systems theorists of the 1960s and 1970s, and he made a distinction between three elements of a system: objects, attributes and relationships. *Objects* are both the actors and parts implicated in the system (and in some accounts constitute the parameters, boundaries or limits of the system). The *attributes* are the properties or features of the objects involved, and the *relationships* are the interactions and communications that exist between the objects. The overall implication is that the system is a 'whole', and the connections between the parts make it a 'system'. Moreover, these connections should be understood and mapped sufficiently to enable decision-making to 'steer' the system. In terms of planning practice, a system may therefore be conceptualised as a town or city, a sub-region perhaps, or something more bounded, such as a transport 'system' or perhaps the location and interaction of retail spaces and consumers. Spatial planning policies and decisions will to some degree influence or shape these components and therefore the system as a whole. Chadwick adds a cautionary note, however, reminding us that:

> like beauty, a system lies in the eye of the beholder, for we can define a system in an infinite number of ways in accordance with our interest and purpose, the world is composed of many sets of relationships. (1978: 42)

This quote begins to reveal one of the main practical issues: the difficulty of placing value on different elements and, indeed, deciding where and how to draw the line between the relevant constituents of the three elements and the system as a 'whole'. The question becomes: What are the limits of the system?

Systems theory therefore implies a degree of boundedness; a closed system which 'limits' or compartmentalises complexity in some way. Ferguson concurs with systems theorists in claiming that variables can be seen as both causal and in producing effects, but arguing that 'you cannot understand a cell, a rat, a brain structure, a family, a culture if you isolate it from its context. Relationship is everything' (1980: 10).

This view begins to unpick the clear distinction between system and environment that some authors emphasise; because some 'systems' are said to be more chaotic and crowded than others and the relations that shape and give meaning to the system are more diffuse. In contrast, the idea of the 'open system' emphasises the multiple and unpredictable flows and relations of systems, such that accurate prediction or control is inherently problematic or risky. Such a notion also paves the way for complexity theory and the stress that some 'systems' are more fluid, unpredictable and may be subject to a range of sometimes unforeseen influences. A prime example of an 'open' system is the weather – where long-range forecasting of the system is unstable and uncertain.

Thus systems thinking over time has tended to characterise the 'system' as somehow distinct from its environment – that it can be in some way isolated. That is, the system can be lifted out for assessment, or lifts itself, from the chaotic or hugely 'noisy' complexity of, say, city life and its day-to-day operation. One implication of this is the criticism that the systems approach tends to simplify the messiness of the world. Deficiencies in this regard may be due to capacity limits or technical restrictions, or may be due to imperfect information, or more darkly, due to some bias or political influence over the input, analysis and then prescription stages when using a systems model.

Systems theory is not a single coherent body of thought but is constituted by a range of different traditions and approaches. In this sense it is paradigmatically diverse, and different features and emphases may be regarded as important depending both on the approach and on the epistemological view prevailing (see Knorr-Cetina, 1999; Kwa, 2002). Taken together this suggests that differing estimations of relevance and scope are likely to be inferred by the different theoretical perspectives which draw on a systems view.

The next idea underpinning the systems view is the idea of *autopoeisis*, which suggests that systems are self-dependent or self-reproducing. In contrast is the notion of *allopoeisis*, whereby the focus of attention is reversed, that is, the unit or system, for example the city, draws in other flows and inputs and is therefore always producing and requiring 'something more' in order to function. In this sense, the system requires 'external' materials or influence as part of its operation. This distinction requires us to accept the idea of a 'closed' system and therefore a wider environment which is deemed to be somehow 'external' to the system, raising important definitional questions about places, cities and economies.

Luhmann's (1986, 1995) work emphasises that autopoietic systems are self-organising and that they may also be self-referential, in the sense that they grow and change in reaction to the actions of the elements 'in' the system. However, systems definitions and operations remain a source of debate among planning theorists (see de Roo and Silva, 2010; Rasch and Wolfe, 2000) and we consider these arguments below.

The flow of 'energy', or the inputs into and through a system, is seen as an important driver of system dynamics. In a planning and development context, a broader conception of 'inputs' into the system is necessary, covering: information, money, labour, discourses and, we venture to suggest, plans and strategies. The flow of such 'energy' helps drive the system, such as a city-region or the property development process, by feeding the 'noise' that sustains the system into distinct states, yet promotes adaptation and prevents entropy. Our use of the word 'noise' here recognises that entropy is often used to refer to a state of disorder or randomness in accounts of complexity. Entropy describes the tendency for systems to either rundown or 'fail' and its counterpart of negentropy refers to the system elaborating and mutating over time. A similar label used for the latter characteristic is that of 'emergence' where, change, communication and feedback act to inform and alter the system.

While systems theorists interested in planning, such as Chadwick (1978) and McLoughlin (1969), were not blind to the difficulties in applying systems theory, it did become quite an important component of (particularly) strategic planning during the 1960–80 period. Subsequent developments in thinking have meant that the label has been supplanted somewhat by complexity theory since the 1990s. However, this interpretation or development is still driven by the need to understand the relations and conditions for successful policy implementation, and has been assisted by the parallel re-emergence of the environmental agenda and its broader discourse of sustainability. Sustainability priorities cut across the range of human–environment relationships and require the type of integrated thinking to which system theorists lay claim. Such analysis is often populated by related ideas such as eco-system dynamics (see World Resources Institute, 2008) and socio-technical systems (Mol and Law, 2002), which recognise the importance of interactions between humans, nature and technology. Yet, despite a new wave of interest and a degree of political awareness of the 'complexity' of sustainability, similar questions over interpretation and measurement persist.

Complexity and Complex Systems Theory

Systems theory applications have tended to assume a degree of equilibrium in the system. This means that the components or actors stabilise or can be viewed as being sufficiently predictable that forecasting and predictions of 'system' behaviour can be reliably made. In some contexts it may be appropriate to use this kind approach when, for example, managing traffic flows or analysing a reasonably limited or 'closed' process. However, when the network of actors and components is more diverse and uncontrollable, it brings with it a concomitant increase in variation and unpredictability. This raises a distinction between 'organised' and 'disorganised' complex systems, where complex systems are:

> Characterised by a large number of actors acting both in sequence and simultaneously. The actions of each unit on a particular historical period however, only depend on the interactions with a subset of those actors. (Byrne, 1998: 20)

As such, the complex whole is not necessarily predictable or bounded. Indeed Cilliers (1998) argues that the (merely) complicated can be analysed completely, and that the extent of the components and their functions can be understood and predicted. In similar vein the organised complex system will be predictable to some practical degree. The range of variables and possibilities present in a complex socio-technical system that is deemed to be disorganised or 'chaotic' is likely to be unquantifiable and less predictable, with influences potentially coming from different sources, locations, actors and times, and with the interactions producing 'emergent' properties. In reference to this, some complexity theorists have it that *relationships* between entities in the system may be aggregated-up to reveal an organism that has 'a life of its own', that is, it is seen as being more than the sum of its parts. This is elaborated by Capra, although here the idea of wholeness is preserved:

> Systems are integrated wholes whose properties cannot be reduced to those of smaller units. Instead of concentrating on basic building blocks or basic substances, the systems approach emphasizes basic principles of organization. Every organism – from the smallest bacterium through the wide range of plants and animals to humans – is an integrated whole and thus a living system. The same aspects of wholeness are exhibited by social systems – such as an anthill, a beehive, or a human family – and by ecosystems that consist

> of a variety of organisms and inanimate matter in mutual interaction. What is preserved in a wilderness area is not individual trees or organisms but a complex web of relationships between them. (1982: 266)

In this account the system is highly networked and the use of 'emergence' is a reference to the reproduction and evolution of the system. As such, efforts to forecast system iteration and to model changes will not always be adequately represented. One view has it that the disorganised system may not be satisfactorily mapped or indeed reduced to suit the researcher's aspirations or aims – although it may be tempting to do so. One option is to try to break down the system interactions or to look at smaller scales; such as the complex system of a particular development (Doak and Karadimitriou, 2007), but still serious questions persist about generating a 'full' understanding of multiple interactions and compound reactions:

> [complex] systems are constituted by such intricate sets of non-linear relationships and feedback loops that only certain aspects of them can be analysed at a time. Moreover, these analyses would always cause distortions. (Cilliers, 1998: 3)

Some complexity theorists contend that complex systems relationships are both dynamic and contingent. Equilibrium in the system becomes a subjective matter and of relative degree, or is otherwise affected by the observer's needs, or by assumptions made in terms of aims and methods. Viewing societies, regions or cities as 'systems' places emphasis on an holistic set of interrelations and interdependencies in which 'planners' and 'planning' are only one of the actors involved in the process of change and adaptation. Although there are different views about the level of control or influence that planners have in this system, there is some consensus that the system can be shaped by them in some way(s). What we then work with in planning terms is an incomplete picture and an incomplete understanding of the nature and importance of some relations and actants (see Hillier, 2007; Rasch and Wolfe, 2000). We now go on to reflect on planning and the assumptions of complex systems theory for the enterprise of planning.

Planning and Complexity

Complexity theory problematises the past application of systems theory and challenges much of traditional planning practice. It cautiously develops a view for planners in which experimentation, adaptation, learning and deliberative practices are required, rather than 'top-down' ideas of system management that have pervaded many planning approaches in the past.

The reticulist component of planning practice tends to 'fit' closely with the worldview painted by complexity theorists. However, there are some disagreements among them regarding the ontological basis of complexity. The relational networks required to formulate and implement spatial planning policies have always been a key concern of planners, and the unpredictable and non-linear impacts of planning decisions have troubled the profession for many years. Similarly, the emergence of new urban forms and spatial arrangements through multi-actor negotiations, structured by a range of political and economic forces and factors has forced planners to recognise their role as partial and, at best, catalytic. It is not surprising, therefore, that planning theorists and researchers have increasingly drawn on complexity theory to conceptualise, analyse and understand the contexts and processes that operate in these spheres. Having said that, some analysts feel that the implications of complexity theory highlight an impossibly fine-grain and multiple world and that responding by either placing emphasis on particular aspects, or simplifying the complex system, is unwise and inherently flawed. Others want to try to work more pragmatically with that imperfection to do the best they can in the circumstances. Kwa (2002) uses the terms 'baroque' and 'romantic' to map the attitudinal continuum and this has been applied in a planning context by Hillier (2005). In simple terms, the 'romantics' believe that it is possible to reduce the complex system to allow for analysis and prescription while 'baroque' thinkers see complexity and 'chaos' everywhere and cannot justify its simplification. The latter reject the idea that a snapshot or a claim to (practical) comprehensiveness can do justice to the emergent and fluid reality of complex systems (for an illustration of this continuum, see de Roo and Silva 2010).

Chettiparamb's work provides a well-grounded appreciation of the range of ideas about complexity and develops an approach that sits towards the 'romantic' end of the continuum. Her study of the People's Planning Campaign (PPC) in India (Chettiparamb, 2007) draws on the theory of autopoiesis to explore and reinterpret the process and outcomes of government reforms in the southern Indian state of Kerala. Taking her cue from Luhmann (1995), she maps out a network of 'coupled systems', based on the following principles (Chettiparamb, 2007: 494):

- that there are three types of autopoietic systems – living, psychic and social – each using life, thoughts and communications respectively as their mode of system reproduction;
- that social systems are formed by the creation of some form of distinction with the environment;
- social systems use functional differentiation based on the formation of a binary code to structure themselves internally;
- there is no access to objective reality: all observers see events through the lens of a particular system;
- the 'organisation' of an entity defines the particular relations that make it a member of a class; its 'structure' defines the actualisation of these particular relationships in particular examples;
- that autopoietic systems maintain organisational closure while remaining interactively open.

Applying this framework of self-referential complex adaptive systems, Chettiparamb explores the institutional restructuring and radical decentralisation that was taken forward during the PPC. A change in planning arrangements was achieved through a loose coupling of the legal, administrative and political systems and by deploying a mixture of legislative, ideological and administrative tools (e.g. central government legislation, the adoption of particular political principles and government orders). This arrangement attempted to 'structure' and stabilise the complexity involved, but allowed room for interpretation, experimentation and co-adaptive learning. She points out that this perspective provides a number of insights for planning practice, including;

> the inevitability of selections, the problem with target-driven control, the necessity of retaining a principle of variation for stability, the importance of multi-level differentiation relating to generalisable dynamics of person, role, action programme and value at various levels, the advantages of self-referential systems and the ability of such systems to be coupled together on projects or events, resulting in different interpretations of events and different processing. (Chettiparamb, 2007: 506)

A more 'baroque' interpretation has been taken in the work of Innes and Booher (1999, 2010). They have been attracted more towards the learning, adaptive and self-organising characteristics displayed in complex systems and have applied these ideas to evaluate processes of consensus building in planning practice. They lay out a set of criteria for evaluating the processes and outcomes of consensus building in planning, which suggest that effective consensus building should:

- be self-organising (allowing participants to decide on ground rules, objectives, tasks, working groups and discussion topics);
- foster creative thinking;
- incorporate high-quality information of many types and assure agreement on its meaning;
- result in learning and change in and beyond the group;
- result in institutions and practices that are flexible and networked, and which permit the community to be more creatively responsive to change and conflict.

They also point out that immediate or 'first order' outcomes might reverberate through the system leading to second and third order effects. For example, innovative practices at project level might lead to wider changes in practice which, in turn, could result in new institutional structures or policy discourses that pervade the system. In the UK, the example of the so-called 'Merton rule' comes to mind, in which a specific policy initiative by the London Borough of Merton (seeking a percentage of on-site renewable energy provision as part of new development schemes) was first copied by other local planning authorities and then taken up by central government. This led to wider programmes of on-site renewable energy provision (such as the facilitation of the management of on-site renewable and national policy change).

These two examples illustrate how complexity theory is being used to analyse and shape planning practice. Its holistic epistemology lends itself to the multi-dimensional and multi-layered probing of dynamic planning processes and the contextualised transfer of principles and lessons between local, regional, national and international planning systems. Yet there are clearly some challenging aspects, both philosophically and practically, which need to be understood and carefully negotiated where complexity theory is used in practice.

Conclusion

It should not be forgotten that systems and complexity are socially constructed views of the world and their 'truths' compete with other views and theories in helping us to understand and act in the world. This also means that they are inherently simplifications of the world, shining light on certain elements and relationships which reflect the intended or unintended prioritisation of planning practitioners or researchers. Nevertheless, this discussion of systems and complex adaptive systems has shown the significance of:

- a longstanding concern with relational dynamics of planning and a realisation of the complexity of social and spatial interaction;
- the ebb and flow of a systemic view of planning, from urban systems to even more holistic socio-ecological-technical systems (brought about partly through the rise of the sustainability agenda);
- the difficulties in conceptualising and operationalising these ideas to support planning practice (e.g. actor identification, boundary definition, system dynamics);
- the utility of drawing on these ideas to (re)understand planning practice and the relational webs it is a part of and has to work through.

Thus complexity theory and the systems perspective add to a 'toolkit' in terms of key ideas and can assist in organising planning, that is: thinking in an integrated way, attempting to anticipate future change, and structuring planning practice towards adaptation, experimentation and learning. This perspective attempts to confront the multiple factors and relations that shape places and associated social, environmental and economic conditions. Many planners already understand that their efforts are often frustrated by multiple and often conflicting factors – complexity theory and systems analysis can help to provide a roadmap, albeit incomplete, for their efforts to plan. As Chettiparamb suggests: 'Complexity is integral to planning today, and the profession needs theories, understandings and mechanisms to deal with it' (2007: 507).

FURTHER READING

Good introductory material on complexity is found in Byrne (1998), who also begins to make reference to complexity and cities, while Battram (1998) provides a nice simple primer in complexity theory. Work by Cilliers (1998) and Capra (2002) has been influential in informing debate over complexity and they discuss the big ideas in an engaging way. In terms of systems theory, Chadwick's book (1978), although dated, is still a very comprehensive and thoughtful overall discussion of systems theory and its use, and possible misuse, in planning. The collection of papers on complexity found in the de Roo and Silva book (2010) highlights the state and range of thinking on complexity in the 2000s. There was also a special issue on complexity published in *Theory, Culture and Society* (2005), which provides useful oversight and critique through theoretical extension and application.

6 HIERARCHY

Related terms: scales; policy cascade, plan hierarchy; decision-making; implementation; control; organisation; subsidiarity; discretion; network

Introduction

Hierarchy is used widely in a range of contexts to express a form of ranking and may imply some organisational command chain used to guide decision-making and implementation. This chapter is concerned with different ideas and examples of hierarchy and the structure and overall constellation of plans and policies, legislation and law that shape 'planning hierarchies'. The term hierarchy surfaces in numerous ways where policies are disseminated and applied in planning practice, as well as being latent in the way that many organisations and bureaucracies operate. Our main purpose is to demonstrate that hierarchies are found and organised in planning, and influence the way that planning processes, decision-making and implementation take place.

Planning activity, particularly at the strategic and national levels, requires that policies and plans are disseminated and implemented locally. This follows the idea that small decisions form part of a larger whole, or help achieve more strategic, national or possibly global objectives. Issues of equity and typical features of territoriality also demand that planning systems are operated fairly and consistently. If national, or even global, policy aspirations are to be implemented locally the question becomes: How does this happen? What organisational structure and decision-making processes are found in planning? The coordination and implementation of policy also links this chapter to questions discussed in Chapter 7 (Implementation) and of course to the general consideration of planning provided in Chapter 2. Our account below highlights how typical understandings of hierarchies tend to be rather idealised and come under intermittent scrutiny and revision.

Definitions and Characteristics of Hierarchy

Social hierarchies are often created and maintained in an attempt to control, regulate or standardise actions and decisions taken, or otherwise to maintain distinction in society. Accounts of organisations and bureaucracy in the policy sciences debate hierarchy and decision-making widely. Hierarchy in terms of organisational theory, management and politics, refers to an ordering and arrangement of a tiered system, or of the component parts in an organisational 'cascade'. This typically features levels or ranks organised with a view to exerting some degree of control or direction to that system. The most common form is a vertical hierarchy whereby power and authority are exerted and delegated downwards, with feedback looped back upwards in some circumstances. The actors maintaining or organising the hierarchy may set out priorities and attach relative degrees of importance to different considerations; they act in this way ostensibly to 'set the agenda'.

Individuals maintain a personal hierarchy of priorities that reflect values, objectives and available options which may govern spending, or the selection of leisure pursuits for example. Such choices are affected by a range of cultural, social and economic influences, which act to shape the individual's attitudes and decision-making. The application of the term hierarchy to personal decision-making is not the main focus here. Rather, we are more interested in first, how planning as a bureaucratic process is orchestrated and secondarily, where other uses of this term surface in planning practice.

There are numerous labels and types of hierarchies which have been identified, each with supposed benefits and drawbacks and which are detailed in the public administration and management literature, and as part of efforts to develop organisational theory (see, for example, Daft, 2009; Thompson et al., 1991). Many organisations have moved to try to 'flatten' hierarchies to encourage interaction, feedback and innovation; in planning and local governance this may be seen in efforts to encourage partnership working and community engagement. Equally the way that such structures operate necessitates an understanding of discretion, bearing in mind Ham and Hill's observation that 'all delegated tasks involve some degree of discretion', highlighting that specific circumstances precipitate variance (1993: 152).

Another significant factor here is the underpinning operating logic that persists in a (planning) system. This may mean that differences in the organisation and the routinised flows up, down or around the hierarchy will vary according to the types of information and knowledge deemed important or legitimate. The point being that planning goals and the evidence to support policies cannot effectively be operationalised without some form of organisation and hierarchy linked to process and decision-making. The way in which the hierarchy is colonised by dominant cultures, the influence of powerful interests, and the management of the hierarchy through rules, resourcing and other regulatory tools all play a part in conditioning and shaping system operation. These considerations also play at the edges of both procedural planning and substantive planning, that is, in both the mechanisms of planning and the aims of planning.

Dean views the operation of the state and attempts to orchestrate actors as involving a wide range of tools and instruments as part of a regulatory environment that includes: 'multiform projects, programmes and plans that attempt to make a difference to the way in which we live' (1996: 211). Such efforts involve not only written and formal policies and laws, such as planning policies, but a whole range of other discursive mechanisms. The degree of complexity and the range of different factors and considerations faced in the lower orders of the hierarchy can also be daunting. Part of the local planners' skill becomes, in this view, an exercise in gathering, condensing, interpreting and applying such influences appropriately: both to the case in hand and in such a way that powerful authors above can reach agreement with policy interpretations and recommendations. Such an array of rules, processes and priorities also produces great difficulty for members of the public to engage with and contest planning.

At least three general features are seen consistently in the operation of hierarchies and which are important here. These are: firstly, the implied *prioritisation* of actions or options; secondly, the notion of *sequencing* or timing of steps or assessments; and thirdly, the imposition of order, or *framing* in terms of attempts to control and standardise actions. The last aspect involves more or less oversight from those operating above in the hierarchy, for example a manager routinely checking a subordinate's work before it can be 'signed off'. These features can also be seen in classic descriptions of 'how to plan', and certainly in terms of regulatory or allocative planning, as described by Glasson: 'allocative planning is concerned with co-ordination, [and] the resolution of conflicts ensuring that the existing system is ticking over efficiently through time in accordance with evolving policies' (1978: 20). As such, a similar situation may be seen where a draft plan will be checked and confirmed (or rejected) by a higher authority.

Indeed the shorthand label of 'planning system' reflects the way that planning decisions are organised with reference to policy and priorities set out at a number of connected levels. There is necessarily some notion of the imposition and acceptance of a set of vertically structured relations implied here, and that these are orchestrated in some way. This involves both a procedural hierarchy and a substantive hierarchy where set steps and tests will be applied to ensure conformity, and where the goals or higher level policies are intended to be pursued at successively lower levels. The implication is also that the power of the levels are not equal by some measure; and that the stages, parts or tiers of the hierarchy draw on different resources and may derive legitimacy from consistency with, or adherence to, policy or other conditions set out above in the hierarchy.

Such structuralist views of hierarchy also stress that an explicit or implicit ranking is central to its operation. In this view the concept of hierarchy may be explained as a form of 'ladder' extending the 'reach' of the state, whereby actions can be dictated centrally and evenly across the territory. Conceptually the arrangement of planning typically forms a policy hierarchy, which has traditionally operated as a top-down policy cascade, where policies are translated and applied at successively more local scales. Self defines this as a: 'positive hierarchy...where firm instructions move down the line' (1977: 69). The national layer is traditionally the most powerful and in Self's terms it is 'authoritative'; stemming from where planning policy has emanated. Yet this 'source' is influenced by tiers or actors from both 'above' and 'below' the level of the national. The upward and downward looking nature of vertical hierarchies also enable certain decisions to be passed up or to remain at an upper level of decision-making:

> the logic of formal authority, which must reside somewhere, compels the harder issues to be shifted upwards to groups of decreasing size. (Self, 1977: 205)

This means that some decision authority is delegated, some retained and some always held at a higher level. This is shown most clearly in the English planning system where larger scale developments may be 'called-in' or referred to the Secretary of State for a decision following a tacit application of the principle of subsidiarity, that is, where decisions are devolved or taken at the level deemed most appropriate. Most often this occurs when a development or policy change can be cast as being of national significance, or where a scheme does not conform to the existing local plan.

How the policies and decisions are made may therefore be seen more as a product of networks rather than through the operation of a vertical hierarchy controlled by one authority. Our view is that a pluralist hybrid has developed in the UK, whereby the discretion and influence of a wide range of actors operating and lobbying at different scales subvert and distort the hierarchy, which acts as a frame to provide overall direction and shape planning decisions. Thus the hierarchy is an attempt to illustrate how things should happen and where authority resides, but there are other interactions and flows that actually shape decisions. Different political ideologies may also wish to maintain or downplay forms of hierarchy and decision-making that suit their philosophy, with more liberal regimes typically seeking a more decentralised approach to decisions.

Hierarchy and Planning

Typically governments use a variety of state machinery or apparatus and attempt to cascade policy aspirations downwards and across a territory using different means. This orthodoxy is replicated widely, and is very clearly seen in planning. Cullingworth and Nadin contend that: 'a plan-led system requires a comprehensive and up-to-date hierarchy of national policy, regional strategies and local plans' (2006: 81). It is also implicit in such statements that vertical integration is deemed to be instrumentally necessary as the legitimacy, viability and legibility of planning policies and goals would be threatened without such a decision tree. This type of hierarchical approach is also typically justified as it provides a degree of transparency. The necessary support is maintained by technical evidence and quantitative forecasts, marshalled by experts as policies effectively 'carry' and apply power downwards. Such processes are rarely smooth or without controversy, as highlighted below.

Hierarchy and planning policy

The classic vertical hierarchy of plans and planning authorities is the most obvious expression of hierarchy in planning with tiers of policy, plans and actors grouped at different scales. Indeed formal land use and spatial planning remains broadly top-down, with national and international policy and legislation shaping the limits and designs of local policy and practice. The way that planning tends to transpire rarely accords to this stylised view, however, and has been subjected to extensive critique in the light of a more fragmented, globalised and networked society and the associated post-structural critique of hierarchies and scale. Thus in post-structural terms, such considerations of hierarchies are being supplanted by the concept of networks. In application to planning, this view of the 'system' reconceptualises planning as comprising a network of relations and actors which is maintained primarily by a policy community at the national level, although with key tools such as plans acting as intermediaries.

Despite critique and dissent over the form and manner of a hierarchical planning, the policy 'hierarchy' persists to assist in legal calculations of legitimacy and procedural transparency. It acts as a structure to aid clarity and provide an indication of intent and overall trajectory. Healey (1990) sees this as being important in clarifying 'decision rules' or criteria to promote a uniformity of policy adoption. The tiers of the hierarchy all produce policies and plans that are handed down for interpretation and broad conformity to be followed by the more localised policy tiers or for other actors, such as developers, community groups or local politicians. This ensures, or it is desired, that the intentions of the higher level planners and policymakers are cascaded downwards. This means that successive policies and plans are arrived at by a process of 'bounded negotiation' where certain options or priorities are largely predetermined by the immediately pre-eminent scale of the hierarchy (and the associated negotiations that led to that product). That said, the planning systems operating in the UK exhibit considerable flexibility and local interpretation in 'filling out' the general requirements of a 'plan-led' approach (Haughton et al., 2009). Such discretion in terms of decision-making and the accepted ability to add or refine policy handed from above is tolerated in order to ensure that local requirements and experiences are given attention. This ensures that the general thrust of policy is more likely to be accepted on the ground.

Ambrose (1986) in his renowned Marxist critique of planning argued that powerful interests with most at stake in the system will have their views reflected in its implementation. This interpretation of planning and hierarchy brings into view the way in which planning is shaped by the political and economic context in which it operates. In this vein it is important to underscore how the planning hierarchy is subject to almost continual revision and contestation as interests 'jostle' to amend, interpret and integrate policies and strategies from 'above' and 'below'. This conceptualisation implicitly rejects the traditional idea of a ladder or vertical hierarchy. It views the relations and relative power of actors as far more fluid and networked, with interests 'jumping scale' and shaping planning policies and decisions opportunistically and as their resources allow.

In these politicised conditions it is hardly surprising that lobbying at national levels (and sub-national levels) is vigorous. Different interests seek to affect the general policy frame and trajectory given that this is where the 'big decisions' are taken over policy goals and targets. Hierarchies are therefore open to influence and challenge. This also includes strategies used by 'local' interests in challenging and negotiating policy handed down from above. Owens shows this in relation to the siting of controversial development schemes and protests which:

> calls into question the concept of the policy 'cascade', whose advocates seek to restrict debate about generic issues at local inquiries. It is argued that local resistance both provides an institutional platform for, and is in turn reinforced by, a wider policy critique. Arrangements for consideration of specific projects therefore provide crucial apertures for debate about national priorities, and repeated controversy acts as an important longer-term stimulus to policy learning and change. (2004: 101).

This type of argumentation has been seen on numerous occasions in planning and there is often the possibility of challenging and reshaping policy from below. Controversies or other feedback from the local or from individual cases can bounce back upwards and reconstitute future policy from above. For example, local protests or a legal decision concerning a particular case can force government to rethink policy, as was the roads policy in the mid-1990s following direct action protests (Seel et al., 2000).

Applications of Hierarchy in Planning Practice

As we have stated, iterations of the planning system in England have exhibited a clear policy hierarchy, which is intentionally created in order to allow for transparency and an effective cascade of general goals and policies. Yet rarely is this smooth or consistent due to: the complexity of information; actors and the commonplace disruption of policy change; and sequencing of policy flows that combine to produce a quite diverse set of plans, messages and outcomes. The argument is that despite such a degree of divergence and confusion there is also a measure of orchestration. What we can say is that the way that policy and process have been found wanting has led to persistent criticism of 'planning' as a burden. We now turn to a brief description of the key components of the planning system in England and how they have been aligned.

The hierarchy of plans and policies in England

In the period 1992–2011, Planning Policy Statements (PPSs) were prepared in England and revised on an ad hoc basis. These have applied across the whole of England (and separate versions apply in Wales, Scotland and Northern Ireland). Regional Spatial Strategies (RSSs) were drawn up under the Planning and Compulsory Purchase Act 2004, and 'below' these sat the local development frameworks (LDFs) or, prior to this, local plans that covered the district level. These three (or four) components were the headline elements of the planning policy hierarchy for England until 2011.

In addition, some general efforts at setting global policy agendas, and also at the European level, impact on planning, such as the Kyoto protocol and the related climate change agenda, which add a 'soft top' to the planning hierarchy. Again, for more detail on the tiers of planning and associated plans and policies, and other instruments used in physical and economic planning, see Cullingworth and Nadin (2006) and subsequent editions. Examples of the key policies and organisational structures that form the 'planning hierarchy' or hierarchy of plans relevant in England as of 2010/11 are:

- Europe – the European Spatial Development Perspective (ESDP) and other EU level directives, see: http://ec.europa.eu/regional_policy/sources/docoffic/official/reports/som_en.htm. This is a very broad and a rather fluid scale of the planning hierarchy but it is bolstered by specific interjections through directives and other European legislation, for example, the Natura 2000 Wildlife directive, see: http://www.defra.gov.uk/wildlife-countryside/ewd/ewd09.htm#euwbhd.

These are drawn down and interpreted at the national level and passed through to regional and local plans and strategies as applicable.

- National – in the UK through various policy instruments but primarily shown in PPSs and circulars, as well as other mechanisms such as government White Papers. In England this policy tier is primarily generated through the DCLG, see: http://www.communities.gov.uk/planningandbuilding/planning/planningpolicyguidance/.
- Regional – through RSSs (explained in *Planning Policy Statement 11* (ODPM 2004d)). This level of planning has been contested and refigured on numerous occasions, the latest being the current coalition Government's dismantling of Labour's regional tier of plans and institutional arrangements. But some arrangement is likely to subsist and operate as part of the hierarchy, even if not providing a comprehensive direction (see Glasson and Marshall, 2007, and see 'Local Enterprise Partnerships').
- Local level – through the Local Planning Authorities in England, in various local plan arrangements. See for example, since 2004, Local Development Frameworks, as explained in Planning Policy Statement 12) (ODPM 2004b, and see also Chapter 4 in Cullingworth and Nadin, 2006).

A diagram of the way that planning is supposed to conform to this type of simple hierarchy was usefully set out in several of the PPSs (see PPS 1, 11 and 12 (extant until 2011)).

At the national level the PPSs are issued and revised from time to time and they cover a range of policy topic areas. These set out the broad policy objectives and guidelines for the way in which planning actions and restrictions should be designed at regional and local levels. There are also other similar policy statements relating to topics such as minerals planning (Minerals Planning Statements (MPSs)). While many of the PPSs focus on particular subjects (such as transport, or housing), three PPSs set out the process and structural design of the planning system. The approach to the hierarchy of plans was explained in PPS 1 and 12, and PPS 9 on RSSs set out a similar set of guidelines for these instruments (until 2011, since when a single National Planning Policy Framework has replaced the system of PPSs; see DCLG 2011).

At each step the evidence base for detailed policy prescriptions is required and external scrutiny and examination are also required. This is designed to ensure that the plans have followed due process and are assembled following the guidelines set out in the relevant legislation and other policy statements. If these conditions are not met to the satisfaction of the relevant Secretary of State, the plans can be rejected and revision required of the local or regional planning authority. In operational terms, the way that such policies at different levels are observed and cited forms an important part of the decision-making process for the approval or refusal of new development.

Sequential testing and hierarchy

Planners also make use of hierarchical decision-making, in particular, when deciding how to prioritise the use of land. Using an implied hierarchy, the principle of 'sequential testing' of land availability is an example of this. The test has developed whereby planners assess whether there are any other sequentially preferable sites to deliver the required facility. This is where the planner seeks to identify, allocate or develop certain types or locations of land before others. For example, using brownfield or previously developed land before greenfield sites, and using town centre sites before out of centre locations are identified. This approach acts to create a hierarchy that prioritises particular sites over others; if no suitable sites are deemed available for a particular type of development then the next site lower down the hierarchy will be considered. The process of following this sequence is often carried out by private sector planners in the UK and they will need to demonstrate that this has been done to the satisfaction of public sector planners and politicians acting as local decision-makers.

In many countries similar general approaches to planning have been adopted with national level objectives forming the key or most influential basis for planning policy (see Alterman, 2001). The power or political will to enforce policy does, however, vary considerably and lobbying groups will try to seek adjustments or exceptions. This takes us towards considering different models of the policy process and how different interests are involved or constrained from shaping policy. It also brings into the account how the reorganisation of policy and governance arrangements affect hierarchy. A short paper discussing how the planning hierarchy operates at and through the emerging city-region (sub-regional) scale is included in Counsell and Haughton (2007). Vertical hierarchies are being reshaped and 'rescaled' as the economy is increasingly globalised and as flows and cross-boundary challenges are faced. This not only shows how policies are supposed to 'fit' together, but also highlights how they are often confused or only partially understood or acknowledged in the context of a 'crowded' policy environment. Thus, the intent of actors to follow hierarchical processes may be imperfect or frustrated. Such a situation may be due to the complex nature of planning and the (in)ability to 'bring it all together' in a way that satisfies national government or other interests. This is exacerbated by the tendency for policies, processes and staff to routinely change and for consistency or continuity to be lost. Therefore variable knowledge, the use of 'discretion' as mentioned, the reality of out-of-sequence timing, and production of different plans and policy iterations, all add to uncertainty and may be viewed as disruptive factors that act to undermine effective or consistent operation of the planning hierarchy. As if this is not enough, the shape and content of plans at the local level can also be undermined by a range of other possible 'material considerations' that affect decisions.

Conclusion: Hierarchy as a Basis for Clarity in Planning?

Policies are set down in a variety of forms and at the national and supranational levels planning brings together numerous strands of policy across a wide range of topic areas. In most jurisdictions there will be a preferred or indicative policy direction and associated priorities that each region or locality will be expected to follow. In order to disseminate and implement policy a formal hierarchy may be constructed. However, the idea of hierarchy can be viewed in two ways: the first is the formal and structured approach towards setting out policy requirements and pushing those downwards. In support of such plans and policies, multiple other circulars, strategies, reports, legislation and policy modifications play a part in shaping the form, content and application of the policy cascade. These will be taken into consideration when formulating or contesting policy, or when deciding or negotiating proposals for development. Adherence to policy intent organised at the apex of the hierarchy is also often enforced by recourse to a range of forms of scrutiny, review and the use of resource triggers. These might be characterised as the 'sticks and carrots' used to manipulate the actions of those at the lower levels of the hierarchy.

The contingency of these bureaucratic iterations has also led to a degree of uncertainty and lack of clarity in terms of what and how to implement planning objectives over time. The second, broader, view of hierarchy when placed in its context includes an appreciation of power and networks and the unequal or asymmetrical influence that can be brought to bear on the operation of the planning system. Moreover, this view highlights how the *content* of hierarchical policy mechanisms has been shaped by knowledgeable and powerful interests. In this way the powerful can influence the nature of the hierarchy (or 'network') itself, resulting in a subversion of the processes followed, selectivity in the information chosen and ultimately the content of the policies produced.

In other planning systems beyond the UK there may be more or less independence or autonomy afforded to local planning regimes. In the USA there is a significant degree of autonomy (see Cullingworth and Caves, 2008). As such the traditional view of hierarchy and planning hierarchies as vertical or as nested tiers of policy remains strong, but should be regarded as a stylised view and can be critiqued as indicated. While hierarchies can be self-perpetuating, they are also contingent and may be more strongly or more weakly enforced, adopted or implemented. They may also be 'hijacked' by particular interests, and any depiction of a clean flow of power and influence from the top towards the bottom may be little more than a gross simplification of the policy process. The planning hierarchy is no closed system, and multiple and complex influences serve to subvert or reorient the decisions and outcomes that the hierarchy oversees. As a result intended efforts to ensure democratic and transparent flows would require continual defence and micro-management. In this context the hierarchy is supported and resourced by guidance, circulars, and explanatory supplements. The implied primary role of public sector planners at the local level is to actively implement and apply the intentions of national and regional policymakers, albeit with some discretion and often conflict between competing interests.

FURTHER READING

For a fuller appreciation of how governments organise and implement policy, see Ham and Hill (1993), and on organisational theory generally and networks Daft (2009), Thompson et al. (2001) and Self (1977) provide good background. The changing nature of local governance arrangements and hierarchies are discussed in Lowndes' work (e.g. Lowndes and Skelcher, 1998). Much work on scale and governance has been conducted in political geography, see Bulkeley (2005), also Brenner (2004) and Swyngedouw (2000), where ways that power, space and scale are used and reformulated through policy are considered. A general view of planning and procedural hierarchy is included in the relevant chapters of Cullingworth and Nadin (2006); and see also the DCLG web site for a view of the range of national policy for England, which of course changes over time. The way that the European dimension influences planning across the European Union (EU) member states is increasingly important, with a number of regulatory requirements featured in the UK (and constituent nations) acting to shape planning systems from EU directives. See for example: www.esprid.org, Faludi (2002) and Healey (2004).

7 IMPLEMENTATION

Related terms: policy process; negotiation; goals; resources; power; networks

Introduction

The concept of implementation is linked with considerations of policy-making, given that the process of planning involves the setting of policy objectives. Through the use of other tools, resources and actions, planning structures and processes should be designed or aligned to achieve those objectives. It seems scarcely necessary to emphasise why planners should be concerned with implementation, given that one of the purposes of planning is quite clearly to guide and direct action and to see plans and policies realised.

In their seminal work on implementation, Pressman and Wildavsky (1973) provided an early definition of implementation as 'a process of interaction between the setting of goals and actions geared to achieving them' (1973: xiv). This highlights how planners need to be aware of and understand policy implementation (i.e. how to ensure that goals are achieved) as well as policy formulation (i.e. goal setting). They also go on to argue that the policymaker must also be able to forge the necessary links in the process or 'chain' to get the desired results. This raises several key questions that students of planning need to reflect on in this connection: how might 'successful' implementation be defined? What is needed to maximise the probability of implementation? What are the obstacles? What are the elements or variables involved? And who are, or should be, involved in shaping and realising policy objectives?

There is a wide literature emanating from the public policy field that considers implementation issues. Indeed the most likely response when discussing implementation will be a lament over a lack of action or slow progress on policy goals. The often cited line from *To a Mouse* by Robert Burns in 1785: 'the best laid plans of mice and men often go awry', reminds us how dislocations between the plan and the action, or outcome, is a longstanding and serious issue. It is necessary also, to remind ourselves that the next line of that same poem: 'And leaves us nothing but grief and pain, for promised joy!'. As such, a consideration of the conditions necessary and the reasons for the so-called 'implementation gap', whereby policies are developed and plans are formed but are not fully actualised is significant. Ensuring that processes are appropriate and that the likelihood of action is maximised should be a high priority for actors involved in planning, and it is certainly a central qualifying point for the consideration of implementation as a key concept here.

Implementation and the Policy Process

The definition of implementation provided by Pressman and Wildavsky (1973) is a rather general one and needs further refinement. It is useful to break down the policy process for more detailed examination. Typically the process can be seen as a series of logical steps, the main ones being policymaking, programme creation and then implementation. This last element centres on putting policy into effect. As such 'policy' can be broadly understood as a statement of intent or a set of goals which are often set out with an associated action plan. Policy formulation is not unique to any given field, or indeed to spatial planners, and many organisations develop frameworks to guide their decision-making processes and associated strategies aimed at delivery. Furthermore, the policy process is influenced by other factors, such as local and national politics, economic conditions and legal constraints. Such influences can mean that the process of policy formulation, as well as implementation, is not straightforward. The policy stages are complex because there are typically a lot of actors or competing interests involved, and numerous other variables and contingencies which are likely to affect or undermine initial policy intentions.

Policies for land use, development and other related 'spatial planning' activities have consistently been put into 'plans' and these can be seen as vehicles for carrying policies and for ensuring that affected parties are aware and can respond to those policies. The plans that are given statutory status tend to have more significance or 'weight'. However, they are not the only vehicles on the road, and planners will make use of strategies of various kinds and a wide range of supplementary policy guidance, such as design guides, development briefs, codes of practice, technical standards/advice, and management plans and related strategies. As we will see, not all affected or influential actors will necessarily be aligned or in agreement with the trajectory of those tools and techniques.

The implementation phase has been viewed as the far end of the policymaking process or 'pipeline'; where policies and plans are taken forward into practice and produce tangible results. Some practising planners will use terms such as 'getting things done' and 'the sharp end' of planning to distinguish it from the more abstract policymaking stages. This is partly a product of the divisions of labour that have arisen in planning organisations. Academics and practitioners have increasingly emphasised that implementation should not be conceptualised in this compartmentalised way (see Barratt and Fudge, 1981 and Hambleton, 1986 for early discussions). Rather, the dynamic is an interactive one in which the multiple interests negotiate development outcomes during the process of 'getting things done'. Policymaking and implementation may be seen as part of an adaptive process in which forms of information and knowledge feedback from planning practice and policy objectives are often contested and (re)negotiated, with power relations playing a central part in shaping both of these.

Certainly there are many reasons why implementation has been given increasing emphasis within debates about planning over the past 30 or so years. Understanding how implementation takes place and the forces that act to shape it and knock things 'off course' can actually help planners to reflect and plan more effectively, and experience of planning and policy failure or poor take-up of policy has heightened planners' awareness. Politicians and planners alike have seen many plans and strategies founder or be ignored. Reflective learning can allow those who engage with planning to follow, adjust or indeed challenge policy trajectories more successfully. When seen in this way, it seems rather axiomatic that a clear and transparent feedback loop can help planners and others involved to more effectively influence the policymaking process.

The wider literature on the policy process indicates that there are numerous factors that may frustrate the implementation of policy. Hill views the policy process as a complex and *political* one, in which there are many actors, that is, politicians, pressure groups, developers, civil servants, publicly employed professionals and others, who see themselves as the 'passive recipients' of policy (and yet who also shape implementation) (2005: 4). Indeed, planners rarely lead in the implementation phases but rely on others, such as developers, financial institutions, landowners, public transport companies, other public agencies or even community groups. Those actors can individually or together hold more power and influence over outcomes than the local planning authority – particularly given that they will have control over required inputs such as land or investment funds.

In exploring this topic we can usefully differentiate between two broad approaches to the understanding of implementation. These are the *policy-centred* and the *action-centred* approaches (Barrett and Fudge, 1983). These approaches demonstrate an understanding of both top-down and structural analysis, as well as more bottom-up and agency-led assessments of the conditions and factors that shape implementation. The traditional and dominant approach towards thinking about implementation, until the 1980s, tended to be the top-down policy-centred view. This perspective emphasises successive refinement and translation of policy into procedures and tasks directed at putting policy into effect. The process is seen as linear and uni-directional with policy being unproblematically taken forward into action. However, various studies have shown how a wide range of macro factors can affect implementation, for example as circumstances or key external conditions change and therefore serve to constrain or facilitate implementation.

This distinction between top-down and bottom-up policy views links to Sabatier's (1986) work which highlights that the alternative action-centred, more bottom-up, perspective problematises the traditional starting point of analysis from the view of a single central decision-maker or policy builder. That focus tends to ignore the range of other actors enmeshed into the policy-action process, and it can screen out competing policy networks and iterations that emerge over time and which affect policy intents and outcomes. Problems do exist with this policy-centred view of implementation. Firstly, policy can come back up from issues and experience on the ground. Policy is also largely constructed with reference to certain dominant concerns or issues and with particular foci or aims in mind. Subsequent policy statements are then refined and 'tweaked' as other issues and subsequent interpretations evolve during processes of implementation and as time passes. Majone and Wildavsky (1978) concluded that it is not policy design, but policy *redesign* that is most common in practice as policies are interpreted and altered with prevailing conditions. Mazmanian and Sabatier also concluded that a more refined approach is possible if we look at iterative stages of formulation, implementation and reformulation, rather than assuming a 'seamless web of evolution' (1989: 24) of policy. Changing circumstances surrounding policy mean that goalposts are liable to move and definitive measurement or evaluation of the realisation of policy goals can be problematic. Unless there is a clear and agreed set of objectives and targets and a sophisticated and determined effort made to monitor and evaluate policy (see Hall, 2002: 232) this is likely.

Secondly, the actors or interests involved or affected by policy processes, and who are needed in some way to help realise implementation, can also routinely undermine policy implementation. Those involved in the process of implementation have their own interests, understandings, time frames and 'ways of doing things' that will not necessarily align with the policy objectives or means envisaged in plans. This shifting set of alliances and related negotiations can be destabilised further by changing conditions (political, economic, environmental, technological, legal or cultural) which may act to undermine 'agreed' policies and implementation regimes. The other, third, aspect that the policy-centred approach tends to ignore is the organisational divisions (both internal and external) in and through which the process takes place. In a 'system' in which any one party does not control all of the resources and powers required to implement policy, they will regularly be dependent on others to deliver outcomes that are (broadly) in line with that policy. This underscores the rise of collaborative planning theory and network applications to planning theory (see Chapter 4) that emphasise the relations and cooperation necessary for effective planning (Innes, 1995).

Mazmanian and Sabatier (1989) usefully outlined six conditions or factors that are required, or that will need to be minimised, for effective implementation to follow:

1 that the policy pursued is clear and consistent;
2 there are transparent linkages between objectives and actions;
3 the appropriate authority is assigned to the policy and it is provided with adequate resources;

4 the managers authorised to pursue the policy are skilled and supportive of the policy;
5 the policy is supported by all stakeholders;
6 the policy is not undermined over time by new policies or changes in socio-economic conditions.

These six factors indicate how difficult it actually is for policies to be comprehensively implemented and alert us to the need for a vigilant policymaker to adapt policy and policy tools to the particular environment in which they operate. The conditions surrounding policy intentions are unlikely to align all of these factors, or to align them consistently. Indeed, even if the first five of the factors listed above can be orchestrated by planners, history tells us that the last factor is clearly beyond the control of any individual interest and will inevitably alter the likelihood, quality or speed of implementation.

McLaughlin (1987) reviewed the uncertainties surrounding policy implementation. He unpacked the implementation process and focused on *relations* between policy and practice. Those examinations generated a number of important lessons: for example that policy cannot always dictate what matters to outcomes at the local level and that incentives and beliefs play an important role in determining local responses. Policy-directed change is ultimately seen as a 'problem of the smallest unit'; meaning that implementation can be frustrated at the very local level or in some contexts by individuals. Effective implementation requires a strategic balance of pressure and support and (even if we avoid issues of power for a moment) implies that bottom-up planning and awareness-raising of policy objectives is as important as expert knowledge and policy formulation from above. This critique of the policy-centred approach also reminds us that both inter-organisational and intra-organisational power struggles are likely to be involved in the day-to-day processes of policy implementation. It also helps explain why the alternative policy-action perspective has increasingly been used to conceptualise and analyse policy implementation as a constrained and negotiative process.

The negotiative perspective emphasises (and arises from) the role of dependent or interdependent organisational relationships. This reflects how planners are likely to enter into negotiation with developers, landowners, politicians, residents, infrastructure agencies and others. The main tools in this process are bargaining and compromise. Therefore a failure to fully achieve policy objectives is more likely, or the realisation of goals is slowed or achieved incrementally. Coalitions of interest will also be sought, rather than the pursuit of total consensus. Various tactics will be used to undermine the arguments and strategies of 'problematic' actors. Ongoing negotiation can produce a 'social order' or 'structure of understanding' between the parties involved (in particular between public sector planners and developers). However, the 'structural constraints' on this form of engagement limit the room for manoeuvre (e.g. legislation, case law, finance, organisational procedures and other accepted norms) but even these 'rules of engagement' can be dynamic.

The view of implementation as a process of action and response emphasises reaction and adaption, rather than a linear, top-down process. Policy is constructed based on past experience and put into operation (which generates reactions), and is then adapted, for example, via policy review or more subtle forms adjustment. This perspective also helps place actors (human and non-human) at the centre of the process and takes account of the various influences that result in varied responses. Such influences include: political structures and accountability mechanisms; socio-economic, political and physical pressures and constraints (e.g. the state of local economies or the influence of the local action groups); and access to resources such as grants, legal powers and information.

In taking this view forward, Ham and Hill (1993) have emphasised that it is necessary to look at organisations as they relate to *other* organisations, with action taking place through a 'networking' process in which inter-organisational relationships become one of the central foci of research and understanding. There is also the question of the 'framing' of planning problems, with the bounding of issues and related considerations shaping the overall process, for example what is deemed 'material' or relevant. This provokes consideration of both action *and* inaction and a questioning about why some issues are dealt with by planning authorities and others are not. Crenson's (1971) study of air pollution in the USA is an example of 'non-decision-making'. In that case, powerful interests and associated ideologies shaped the policy agenda in significant ways. The outcome was a neglect of policy through inaction in terms of dealing with harmful pollution.

Tying together and harmonising top-down and bottom-up policymaking and agenda-setting may hold some of the keys to enhancing successful implementation. There is a plausible link here to collaborative planning and capacity building, whereby negotiation and consensus building over policy aims and their appropriate actions are brokered.

The premise is that actors are more likely to agree and cooperate with policy goals if they have been involved or understand the desirability of those policies. Olsson, in this vein, argues that those 'able to create their own interaction rules are often successful in achieving efficient collective outcomes' (2009: 268). This implies that a degree of empowerment and co-design of policy and policy instruments may assist in more effective implementation or policy take-up.

Finally, the action-centred perspective requires us to understand people's actions in the context of their values, ideologies and perceptions (e.g. the influence of social class, professional training, political ideology, religious faith, gender and special interests). This perspective takes us into questions of culture and social normalisation, and has been considered by, for example, Hillier and Rooksby (2002) and Howe and Langdon (2002), drawing on the work of authors such as Pierre Bourdieu. Thus, issues, policies and implementation processes are inherently 'socially constructed' by actors who negotiate the webs of policy and action through a set of network processes. Overall, the difficulties encountered by policymakers and the messy or incomplete process of implementation are unsurprising given the range of obstacles, gaps and 'other drivers'. To draw together our metaphor, we would suggest that the road to effective policy implementation is potholed, roadblocks frustrate us at every turn and there are a lot of rather inconsiderate drivers to contend with.

Planning Practice and Implementation

After the surge of theorising about policy implementation during the 1970s and 1980s, a number of subsequent empirical-based studies explored such ideas in relation to policy implementation dynamics and applied them to different aspects of planning practice. These collectively illustrated the importance of examining relationships *in practice*, which has often involved mapping out the myriad and often subtle interrelationships between plans, policies, institutions, actors and structural forces that go to make up and shape implementation processes. Equally, new theoretical approaches have been adopted by some in seeking to understand the factors that shape policy and implementation processes. These studies have left us with a picture of implementation that indicates that numerous factors act to shape and frustrate policy implementation. This knowledge indicates the range and mutability of actors, values, attitudes, and existing resources, conditions and other policy priorities.

In order to illustrate some of the issues and to provide a 'live' example of an ongoing policy issue, we highlight here a major policy goal in planning that has been beset with implementation difficulties. The policy area of planning for land value capture in the UK, and attempts that have been made to implement this policy, is part of the wider land value question and specifically involves attempts to recoup 'betterment' value from development projects (Grant 1999). There has been a broad policy aim since 1947 (and even before then) to try to recoup at least some of the value uplift that occurs as a product of societal preferences for land use, as channelled through planning policy and the operation of planning consents. When planning consent is granted for development a surplus or 'unearned' value is created. Without state action or policy to address this situation, the value will be absorbed by the landowner and developer as profit. Given that the value uplift is not derived from action on the part of the landowner or developer, governments have recognised this as an anomaly to be corrected or an opportunity to be grasped, or both.

Consequently, various forms of development tax and land value capture have been debated or initiated by UK governments to try to recoup at least a proportion of that betterment value. Many taxes, levies and charges have been discussed or employed, but these tools have proved politically unpopular or contentious, unsuccessful or have been abandoned, resulting in a rather tortuous history where governments have continuously wrestled with the question rather unsuccessfully. This story helps demonstrate some of the difficulties experienced in the implementation of policy and the way in which one policy aim may conflict with other aims or objectives.

As part of the establishment of the land use planning system in 1947, it was recognised that an active land policy would be needed to accompany the new legislative framework. An important part of this approach involved the creation of a development charge (although similar measures had been enabled as far back as in the Town Planning Act 1925, see Cherry, 1974; and Goodchild and Munton, 1985 for a brief discussion of the successive policy tools). The 1947 charge was set up so that the difference in value would be taxed at 100%. This approach was abolished in 1953 after landowners held back land from the market due to their reluctance to sell it at existing use value. In 1967 a betterment levy was introduced and this too failed for similar reasons as the development charge, as well as because the approach was deemed too torturous and complicated to administer effectively. In the 1970s, two further taxes were introduced: the 1973 Development Gains Tax, which was set at 100% of the value uplift and subsequently reduced to 80% by the 1976 Development Land Tax.

The tax-based approach was largely dropped in 1984 by the Thatcher administration and replaced (through the stealth of local planning authorities) by a negotiated, and therefore inconsistent, system of 'planning gain' contributions from developers. The attempts to tax betterment overtly had been abandoned due to a combination of ideological principles (i.e. not to 'burden' market players) and fears that such a tax might undermine efforts to regenerate towns and cities. This 'gap', combined with cut-backs in public expenditure, gave impetus to the much wider use of 'planning gain agreements' negotiated with developers. This was a little-used 'implementation tool' which had actually been available since the late 1950s. This allowed local planning authorities to negotiate contributions (and other undertakings) deemed necessary to 'round out' development and to ensure that impacts and associated costs would be met, or partially met, by the developers (see Chapter 16). This approach has not kept all of the stakeholders happy but it has proved durable. It can be seen as part of a political shift in approach – away from creating a tax based on the principle of recouping betterment value toward a more pragmatic approach. Since the 1990s attempts to refine or replace this system of planning agreements have not been successful. While this tool and the wider issue has persisted, discussions and proposals for various tariffs, charges and levies have left many observers somewhat bewildered (see Barclay, 2010). In 2010 the Community Infrastructure Levy (CIL) was introduced as a voluntary measure to lever in developer contributions, but with doubts about its long-term future and viability in a market downturn (see DCLG, 2008). The 2010 regulations supporting this mechanism can be viewed at: http://www.legislation.gov.uk/uksi/2010/948/contents/made).

The account of so many attempts to tax betterment, or to negotiate contributions to fund infrastructure, is a sobering one and may be seen as a rather extreme case. It will be necessary to read more on the story of this area of policy, as it highlights how the necessary conditions for successful implementation have not been met or have not been stable enough. It also shows that progress towards policy implementation occurs despite obstacles, often as part of an ongoing process of policy (re)formulation and implementation and discussions with key stakeholders. Betterment recoupment has also suffered from the regular boom and bust cycle of the property market in the UK, which means that the viability of many of the approaches proposed has been undermined. This illustrates the significance of the economic context and the power of market interests. Central government have not been able to broker agreement among all the (key) interests and many of the attempted initiatives have proved impractical. There have been technical issues regarding when to charge, how to calculate charges and for what the monies raised should legitimately be used.

In short, the issue has not been satisfactorily resolved and a lack of consensus over the issue persists. Local authorities want to ensure that developments are acceptable in all respects and they want projects to 'pay their way', while developers want to keep their costs down and maintain as much flexibility over the development as possible. The practicability of betterment taxes fluctuates; economic circumstances serve to undermine the viability of the development or can make planning authorities appear not to have drawn out as much of the value uplift as they could (or should?). Landowners and developers can and do withdraw from development when the environment is unsuitable, thus also affecting wider policy aims, such as providing needed housing. In short, the policy over recouping betterment or funding infrastructure in the ways described has not been able to satisfy any of the six criteria set out by Mazmanian and Sabatier (1989) above. Yet, the relative success of planning agreements in some circumstances has rested on their flexibility and the fact that developers and landowners have effectively escaped a comprehensive recoupment of betterment. They have been able to use expert knowledge to invariably pay less than they would otherwise have done under a form of development taxation. Overall, the variable effectiveness of betterment recoupment policies and related implementation tools can be analysed and understood reflecting on the range of considerations highlighted in the policy-action literature.

FURTHER READING

The topic of policy implementation extends far beyond town and country planning and valuable sources of literature are found in the wider public policy and public administration fields. The work of Sabatier (1986) and Mazmanian and Sabatier (1989) still retains generally significant lessons in general terms. Ham and Hill's (1993) work has become something of a classic, taking a broadly political economy perspective to the policy process. The seminal work of Wildavsky is still required reading (e.g. Majone and Wildavsky, 1978; Pressman and Wildavsky, 1973). There was a surge of interest in this subject in planning the 1970s and 1980s (e.g. Barratt and Fudge, 1981; Hambleton, 1986; Healey et al., 1988). Studies looking at planning and implementation cases also include research by Cockburn (1977), Saunders (1983), Cloke (1987), Healey et al. (1988), Brindley et al. (1989), Adams and Watkins (2002), and Murdoch and Abram (2002). Flyvbjerg's (1998) work shows us a case of where planning policy can be frustrated and Greed (1996) looks at planning and implementation more specifically.

8 DESIGNATIONS

Related terms: zoning; boundary drawing; area-based initiatives; spatial segregation; designated areas; special purpose organisations; criteria-based planning; hierarchy; territorialisation

Introduction

Designation and the associated practice of boundary drawing is important in planning practice as it is an indication of how planners and others typically seek to demarcate and bound both space and activities by a process of separation and ordering. Designations are a tool used by planners to highlight particular areas or issues, and are used to help intervene in how particular places, features and activities are managed. Such an approach may be deployed to incentivise certain practices through the creation of designated areas, zones or other targeted policies. In this sense planners act to differentiate between places and spaces so that policies, regulations and funds can be organised on a 'rational' (and often rationed) basis to provide agreed outcomes, or so that specific land uses can be promoted or discouraged. This process is usually justified by recourse to particular strategic policy aims and may be accompanied by more or less specific objectives and guided by criteria.

Designations act as signifiers, indicating the special qualities or needs of the designated area. These designations often reflect a top-down instrumental and rational approach to planning, whereby space is demarcated and separated to assist in organising and administering those areas. This process of delimiting and designating has important consequences in terms of spatial planning, where a key role is to provide guidance for a range of stakeholders, including developers, landowners and local communities. This approach inevitably means that a degree of determinism is involved in setting boundaries and drawing 'lines on maps'. The concept parallels that of hierarchy (Chapter 6), given that it involves treating one place as somehow 'different' or special compared with another place. In other words, this concept reflects a particular subjectivity and can imply a hierarchical approach to place and its ordering, for example particularly in the case of protected areas such as National Parks (NPs), or Conservation Areas (CAs) (see Chapter 18).

What are designations?

A planning designation may be defined for a set period or in perpetuity and is usually set out over a given spatially bounded area. The purpose of designation may be to explicitly set priorities or indicate particular issues to be addressed in that bounded area. The designation may relate to certain agreed positive actions or to a need to restrict particular activity for given reasons. Similarly it may be created to tackle a particular issue such as poor economic performance – enterprise zones are one example (see below) – or to maintain certain continuities, for example architectural or historical character in CAs or landscape quality in protected areas such as NPs and Areas of Outstanding Natural Beauty (AONBs) (see below and Cullingworth and Nadin, 2006; Gallent et al., 2008, for a description of these designations). Criteria may be set concerning whether or not certain activities or outcomes are likely to conform or militate against the aims of the designation. In this sense the designation may be made to reflect the presence of desired features, or it may be made to indicate an absence or other perceived failing. Designations may be broad or narrow and focused on small areas or set out at larger scales.

Why do planners designate?

Designations can act as more or less authoritative or overarching statements of priority about an area, and they have long been used to demarcate and separate out different areas and issues.. As mentioned above, they may be focused, settling only one aspect of the place/area in question or be more general. There are numerous types and purposes for designations in the wide sphere of planning. Thus, designation as a concept involves the means by which planning policies can be applied on an area or a thematic basis, with a view to special provision and/or resourcing. Here we concentrate on spatial designation, but even within this there is often a prioritisation of resources or powers on an issue-by-issue basis.

The process of designation is centred on the way in which particular issues and areas – often interlinked – are identified and isolated for special attention or to provide a signal to developers and others about the way that planning authorities and governments view the future development of that area. Often such efforts are accompanied by cartographical representations that set down the limits of the designated area and to signify the operation of a different regime to the public and other interested parties. For example, in some cases to trigger funding from regional, national or supranational funds – such as European structural funds (Bachtler and Turok, 1997). Designations may also be statutory or non-statutory – acting as codifications of space (as with certain zoning regimes) or, less forcefully, for guidance purposes.

Given that, on a theoretical level, designation promotes or differentiates particular places or activities over others, it acts to create a 'hierarchy of place' (Chapter 6) and re-territorialises space, where a formal ordering of priorities is set down (Foucault, 1970; Murdoch, 1997; Sack, 1986). This demarcation is justified to highlight needs and features that are seen to require policy intervention and to make sure that designated areas are known to the range of stakeholders affected; however, this can also lead to unintended consequences, such as deterring or encouraging visitors or investors with knock-on economic, social or environmental impacts (e.g. Howard, 2004). Partly in recognition of this and in order to fulfil the designation's objectives, they may be accompanied by special funding and grant streams to organise and encourage the associated aims to be fulfilled. In some cases the inspiration for the designation may be a particular political or ideological underpinning, or a perceived need to instigate a looser or stronger planning regime, as with the establishment of simplified planning zones (SPZs) and enterprise zones (EZs) in the 1980s in England, which was linked with national efforts to revitalise certain local economies (see Allmendinger, 1997; Beck, 2001; Potter and Moore, 2000; Thornley, 1991). Another motive underpinning some designations is to help allocate scarce resources to those areas which can use the funds or support to best effect.

Apart from the unintended externality effects that designations can give rise to, another aspect of their 'dark side' is apparent when they are 'captured' or used by certain groups to further their own interests (see, for instance, Scott and Bullen, 2004). A well-established concept in the planning literature is that of 'exclusionary zoning' by which certain residential communities are able to exclude other groups from their areas by utilising planning policy designations or, more correctly, the market processes *around* these planning designations (e.g. CAs, low density housing, green belt or ecological areas) to price out those who cannot afford to live there. Scott and Bullen (2004) also point out that the process by which many designations are identified, and indeed how the governance regime operates, has tended to involve a limited range of stakeholders and has been dominated by elite 'experts' who define the terms and criteria applied. Over recent years some effort to shake off this image and work more collaboratively has been attempted, using principles of good governance and sustainability (see Chapter 3) and in recognition of the wider ramifications of conservation designations (cf. West et al., 2006).

A good example of the 'special arrangements' that planning designation can generate is the NPs that exist in many countries. In the case of the English NPs, the powers and administrative arrangements are different from those for other surrounding areas. NP authorities hold their own planning powers, despite in most cases the NPs cutting across several administrative (local planning authority) boundaries. In this way NPs enjoy a special place in the planning hierarchy. The enabling legislation passed in 1949 (and modified in 1990), also places special emphasis on landscape protection and public enjoyment of those areas (Blunden and Curry, 1989; Gallent et al., 2008).

Zoning is one of the clearest and widely used of planning 'designation' examples, along with designations aimed to protect certain features or landscapes such as NPs noted above. This process clearly demarcates what land use can and cannot be permitted in given areas, with policies written to explain and indicate the circumstances whereby development of different types might be allowed. In effect, the different areas are zoned, for example as predominantly employment land areas, housing, retail and so on. However, on a country-by-country basis, designations of this type may have statutory force, or be part of other non-statutory attempts to shape development and wider decision-making. Therefore, designations may have more or less influence or enjoy authoritative status, with some derived from European or other supranational authorities.

The counterpart to the spatial or area-based approach is often referred to as criteria-based planning where decisions are made or actions initiated when and if certain conditions are met or transgressed. Area-based initiatives (ABIs) involve designation and operate under certain criteria. Examples include where permission for development would be granted if certain preordained conditions are met or a 'check list' is satisfied. This approach is more common in the USA and some European countries (see Carmona and Sieh, 2004). Such designations tend to be made loosely over a wider area and action or funding for particular reasons (e.g. economic development, social disadvantage) is triggered if the relevant conditions or criteria are met.

Planning Designations in Practice

As intimated, the use of designations has a long history in planning terms and they are diverse in nature. We provide a brief glimpse of several examples of designated areas and their purpose, to underscore the range and importance of the demarcation of different spaces and issues tackled in planning by using designations.

National Parks and Areas of Outstanding Natural Beauty in England

NPs have a history stretching back to the nineteenth century with the USA being something of a pioneer in this project with the designation of the Yellowstone National Park in 1872. In England and Wales, NPs were enabled under the National Parks and Access to the Countryside Act 1949, and have a twin purpose of landscape protection and public enjoyment, with a third objective relating to socio-economic welfare added as recently as 1995. The parks effectively separate themselves from surrounding areas as they are deemed to be of special value for reasons of landscape and natural beauty. In the UK, a long history of conservation lobbying eventually persuaded government to create such areas (and also similar AONBs). NPs are drawn up on the basis of presence of special features or landscape types. England, Wales and now Scotland all have NPs and since 2000 two new NPs have been announced in England. The one in the South Downs prompted a rather drawn-out process, with controversy stirred over the boundaries of the new park area. The other established for the New Forest raised local concerns that the NP status would attract an unsustainable level of tourist and leisure use and could cost local tax payers more. NPs have a shorter history elsewhere and exhibit different characteristics, for example land ownership being in public hands in some countries and exhibiting varying administrative structures, with a range of resources and powers being available to them (for more on NPs see below, and also Blunden and Curry, 1989; Gallent et al., 2008; Parker and Ravenscroft, 1999; Thompson, 2005). In the UK NP authorities have their own planning powers and policies, which are expected to prioritise the stated objectives of NPs under the relevant legislation. It is also the case that NPs and other designated areas may receive preferred or prioritised status for grants and project funds, for example in this case for landscape enhancement or other desirable land management practices. The international body that acts as a voice for NPs and protected areas worldwide is the International Union for Conservation of Nature (IUCN; see link below). Often NPs will also have other designations and areas of special status being applied within their area, for example special habitat areas or heritage sites may exist within NPs or wider protected areas, such that multiple and overlapping or a nesting of designations can occur. Often the administrative structure of particular countries, institutionalisation of conservation and environmental protection internationally, and indeed the increasingly supranational process of designating areas, have increased the likelihood of multiple designations existing over the same areas. The clearest example of this is with the European designation of Special Protection Areas (SPAs) and Special Areas of Conservation (SACs) (see http://www.defra.gov.uk/wildlife-countryside/ewd/ewd09.htm and http://www.jncc.gov.uk/page-162). Such uniform designations made across the European Union (EU) are intended to standardise management, improve clarity and raise standards of protection across Europe as a whole.

AONBs are statutory area-wide designations that indicate a particular land area has certain valuable or special landscape qualities in similar fashion to NPs. There were 38 of these areas in England and Wales covering something like 14% of the land area by 2010. Certain activities and development are treated differently within the AONB boundary with landscape protection being a primary concern. Difficulties with achieving the objectives or of ensuring that management plan policies are adhered to are made more difficult as AONBs do not have their own planning authority powers and are reliant on (often a multiplicity of) constituent authorities to adopt and apply the policies of the AONB (see Dietz et al., 2003; Scott and Bullen, 2004), this obviously differs from NPs as discussed above. In recognition of this rather fragmented control, each AONB has had to have a management plan since 2000 which should be signed up to by all constituent authorities through a partnership board or joint committee. The basis for action and policy setting is derived from this management plan, which has been subjected to sustainability appraisal, to ensure that the policies included do not conflict with the key objectives of landscape protection and enhancement, along with the secondary function of promoting public enjoyment that AONBs hold.

There are undoubted difficulties with such designations: how many are justified or manageable? Are they able to be resourced properly? Do they deliver what they promise and are there alternatives? It may be argued that AONBs might be seen as a form of exclusionary zoning with knock-on social and economic implications (Ghimire and Pimbert, 1997) and exhibit some important limitations given their landscape focus. This is perhaps particularly so when rural areas are struggling to reconcile the challenge of integrating the different elements of sustainability and create a multi-functional countryside. Selman (2009) is one of the few voices to question the effectiveness of such conservation designations in the UK and this particular method of regulating space (see also Thompson, 2005, on NPs in this respect).

World Heritage Sites

UNESCO is the world body charged with the responsibility for identifying the most important cultural and natural sites across the world, and in partnership with national authorities, it designates these so that they can be protected and managed sustainably. In practice most sites will already have had some form of national recognition and enjoy a protective designation status. Others, due to their location and the resources of their national governments, may be less well looked after. The World Heritage Site (WHS) designation represents an effort to draw attention to the need for active management of such sites. There were 890 WHSs by 2009 and the number is still growing. The project to identify and designate in this way has proved to be popular with many countries which see the benefits of this 'blue riband' status, including an economic spin-off value in terms of tourism (Graham, 2002; UNESCO, 2010). This issue is common with the NP phenomena and brings similar management challenges, including the multiple designation issue. For example, the stone circles of Stonehenge and Avebury in southern England are a WHS and are partly within an AONB (the North Wessex Downs see: http://www.northwessexdowns.org.uk). They are also scheduled ancient monuments and are within the 2008–2013 EU LEADER programme area aimed at economic development.

Greenbelts

A different form of designation is found with greenbelts. This is a well-known form of designation that aims to prevent urban sprawl and the conglomeration of towns and cities, which has been tried in numerous countries with variable success. The designated area is usually around a city or between settlements, and in that 'belt', development will usually be very limited or restricted. In the case of greenbelts in the UK, development will only usually be permitted in 'special circumstances' (see Elson, 1986; Shaw, 2007); and they have enjoyed their own specific planning policy guidance in England (*Planning Policy Guidance Note* (PPG) 2 (ODPM 2001)) in various ways since 1955. While greenbelts emanate from national policy they are actually local designations with specific greenbelt policies resting in local plans, with the greenbelt boundaries being set at the local planning authority level. As a result, greenbelts can expand and contract with each iteration of a plan (see Amati, 2008; Shaw, 2007). In Japan, when greenbelt policy was applied, notably around Tokyo, the approach failed and was eventually abandoned as the powers and political will to maintain it as a growth restrictor were not present (see Amati and Parker, 2007; Parker and Amati, 2009; Sorensen, 2002). A special issue of the *Journal of Environmental Planning and Management* (volume 50, number 5), concerning recent debates over greenbelt policy, was produced in 2007 (see Shaw, 2007) in response to increasing attacks on the greenbelt concept and in the light of alternative approaches to growth management and the use of urban fringe areas. Amati (2008) gives an international review of greenbelt approaches highlighting their popularity and somewhat variable success.

Conservation areas

CAs were initiated in the UK in 1967 under the UK Civic Amenity Act and by 2010 there were around 9,300 of these in England (English Heritage, 2005, 2010). The aims of a CA is to protect on an area basis the character or amenity (see Chapter 18) of the designated area. The status of CAs will mean that special conditions and extra or more restrictive planning policies will apply to properties within the area.

Typically a neighbourhood or group of streets will form the basis of a CA designation, or in a village setting the whole of the historic core. This designation recognises how the architectural, historical and cultural merits of a place may be 'more than the sum of its parts' so that allowing any of the parts to be removed or altered could detrimentally affect the character and value of the whole (English Heritage, 2001). This approach has been criticised for being overly restrictive and for preventing any organic change in those areas (see Delafons, 1997; Larkham 1996). However, alternatives to such an approach rest with more codified systems of development control found in a number of countries worldwide. Another issue sometimes raised about the CA system in the UK is that it can act to preserve culturally important neighbourhoods tending to benefit existing (and probably affluent) residents and interest groups, creating protected elite spaces.

Enterprise zones

The EZ designation involves signifying special conditions for economic activity in a given bounded area. EZs were introduced in the UK in 1980, featuring a simpler planning regime and offering tax breaks for incoming businesses. EZs were designated by national government to encourage industrial and commercial activity, usually in economically depressed areas such as the Isle of Dogs in London's Docklands area, although they were used right across the UK. Investment was attracted by means of tax reductions and other financial incentives. The zones were intended to be temporary designations, and there have been three 'waves' of EZs. Most were designated in the early 1980s and had a lifespan of 10 years (see Potter and Moore, 2000), with the UK government launching the third wave of EZs in 2011. Similar efforts have been tried elsewhere with variations on the theme of providing special status and different powers in such zones, particularly in terms of inducing an attractive business environment (see Hansen, 1991; Tait and Jensen, 2007). EZs have been criticised for disadvantaging those located outside of the designated area and also of creating a migration of business activity from elsewhere, without necessarily creating net new employment or economic growth. However, the counter view would have it that the targeted areas would not have been lifted at all without such 'special measures'. As such and despite evaluations that are inconclusive about the effectiveness of EZs, such approaches have remained popular with numerous governments.

Chapter 8

Conclusion: Issues and Difficulties with Designation

While designations aim to separate places and highlight issues and priorities in those areas, there are critics of the approach. The designation process has been problematised for a number of reasons. Often the resources available for proper implementation of objectives, and the integration of stakeholders across designated areas has been limited and problematic. This has been particularly acute where no special-purpose organisation or single authority has been in place to operationalise and champion the objectives of the designated area. NPs in England have their own authorities with budgets and act as the single planning authority over the park area. These are often seen as a success in terms of the organisation and management of development in their area and supporting the aims of the NPs.

Other unforeseen or contrary outcomes that may arise from designation include questions over social equity and environmental justice (Bowen et al., 1995) stemming from disparities between places that are and are not designated. Examples are where NPs can suffer from unsustainable visitor pressure and in EZs where a 'depression effect' may be seen in areas around the zones. Critics of designation also cite complexity, cost and the existence of other authorities and powers that could fulfil the aims of the designated areas. Designating areas remains a key element of planning practice, and their use can be seen as reflecting the politicised nature of planning and the tensions between national and local levels of government, for example concerning resource allocations and the perceived ability of local authorities to deliver on national policy.

Designated areas act to highlight specific needs and issues and can act to temper general policy. Designation also reflects a spatial hierarchy of place imposed or constructed across territories and highlighting a range of priorities and issues. In many cases it is likely that special-purpose organisations or associated funding will be necessary to ensure that the objectives of the designation are achieved. Despite problems and unintended consequences, designation can act to bridge between the twin dangers of assuming uniformity of place and the needs of individualised or site-specific responses, which may be needed to deal with the special needs, conditions and characteristics of places.

FURTHER READING

There are numerous accounts of particular designated areas. A number of organisations actively manage or promote designated areas of different types:

- World Heritage Sites, UNESCO: http://whc.unesco.org/
- Association of National Park Authorities: www.nationalparks.gov.uk
- Areas of Outstanding Natural Beauty: www.aonb.org.uk
- Greenbelts: http://www.naturenet.net/status/greenbelt.html
- Enterprise zones: http://www.communities.gov.uk/archived/general-content/citiesandregions/finalevaluation/
- Conservation Areas/English Heritage: http://www.english-heritage.org.uk/server/show/nav.1063
- IUCN protected areas:http://cms.iucn.org/about/work/programmes/protected_areas/index.cfm

Many of the designations are discussed in brief by Cullingworth and Nadin (2006). A critical viewpoint is provided by Selman (2009), and other relevant references for specific designations such as EZs are cited in the text. Tait and Jensen (2007) and Potter and Moore (2000) provide good overviews on EZs. Further key references and readings that include mention of designated areas such as NPs can be found in Gallent et al. (2008). Blunden and Curry (1989) provide the definitive history of NPs and AONB creation (see also Chapter 18). Amati (2008) is useful for reading about the efforts to use greenbelts as an urban management tool.

9 INTERESTS AND THE PUBLIC INTEREST

Related terms: stakeholders; consensus; rights; conflict; power; elites; community; national interest; representation; participation; exclusion; public goods

Introduction

This chapter focuses on the notion of interests and more specifically the idea of the public interest as a term that has been used to justify planning practice and intervention. Given the wide rhetorical use of the 'public interest' and long history in providing a legitimating concept for planning (see Friedmann, 1973; Howe, 1992; Klosterman, 1985), it is important to explain its meaning and usages. Booth, for example, called for 'a wider debate on the nature of the public interest expressed in the land-use planning system' (2002: 169), in order to clarify and make more transparent the thinking and rationale for institutional design and decision-making in planning.

There are different conceptualisations of this idea however, as well as variable applications by different competing interests and as Campbell and Marshall contend; 'it is a term which has often been used to mystify rather than clarify' (2000: 309). It has also been used somewhat flexibly, and at times rather lazily, to interpret different situations and circumstances (Cullingworth and Nadin, 2006). A changing socio-cultural context and periodic ideological attacks on land use planning mean that conceptions of the public interest have been subject to significant amounts of scrutiny and defence (see Alexander, 2002a,b; Campbell and

Marshall, 2000, 2002). Arising from this, it is important to outline how different interests (and 'publics') are involved or implicated in planning and how they are considered, shaped or possibly even neglected.

The Idea of Interests

The idea of an interest may be viewed objectively as being remote from the individual. In this view, interests can be represented without knowledge of individual views or attitudes. Other conceptualisations take a more subjectivised view, seeing interests as individual and partisan. This simple binary has been supplanted by the idea that politicians and civil servants develop communication, understanding and attitudes that are informed by numerous individuals or groups, and resultant views or positions are assemblages of knowledge and influence. Attitudes and positions adopted by decision-makers are sometimes referred to as 'centred subjectivities' where this subjectivity varies with the power and ability of those interests to articulate and to access decision-makers. It is also the case that emphases or interpretations of policy in any given situation or time will reflect the attitudes of government, extant legal provisions and citizen rights.

Planners have arbitrated in allocating resources between interests, ostensibly to maintain efficient economic and social, as well as environmental, conditions. This has prompted some to say that planning itself is largely concerned with dispute resolution and the management of property rights. Planning becomes implicated in the power differentials between interests and the use of the term 'public interest' is as much political as technical. It may be used to justify a rather utilitarian outcome, for example some anticipation of 'doing good' in an often unspecified or rather opaque way, which will be somewhat dependent on which active interests can lever the most support and form convincing arguments. Such situated inequality led to critiques of planning processes and outcomes, for example from Davidoff (1965), who promoted an advocacy model of planning when it was clear that planning processes and outcomes often disadvantaged particular groups in society. This in turn spawned the development of communitarian and collaborative planning models which rose to prominence in the 1990s with their more inclusive and deliberative approach. Campbell and Marshall (2000) indicate how such a collaborative approach has been quite widely welcomed, but that its focus on process is not enough on its own. They argue that a basis for action in the 'public interest' also needs clearly articulated values and aspirations.

Interests in Planning

Planning processes reflect wider struggles and contestations in society. Individuals or groups may benefit directly from planning decisions or policies, while others may perceive or anticipate some form of loss. There will be perceptions of positive and negative impacts for particular interests at different times too, and at different scales through which the operation of planning decisions acts to shape the world. The impacts may be environmental, economic or otherwise affect social conditions or cultural practices. They may be immediate or longer term. Part of the planner's role is to assess and understand such impacts, acting as an arbitrator and a judge about the balance of costs and benefits of development and other planning related activity. As such planning has been predicated and justified on the basis of weighing up such costs and benefits and ensuring 'efficient' land use, while attempting to preserve and enhance, *inter alia*, amenity (Chapter 18) and to mitigate or prevent negative externalities (Chapter 16). In doing this, planners have claimed to act in the 'public interest'; yet this is a rather vague and difficult-to-measure rhetorical justification which requires unpacking.

Lindblom (1959) used the term 'mutual adjustment' and 'satisficing' to reflect how planners were not apparently acting in any one interest, but instead were acting to accommodate and integrate numerous interests. In extending this, Bolan (1983) identified 11 interest positions by which a planner may be influenced. He termed these 'moral communities of obligation', which we group into: 'self-interest', 'professional interest', local authority/'local political interest', 'developer interest', 'local community interest', the 'national interest', the 'environmental interest' and the 'economic interest'. All of these are likely to overlap or be in tension in different ways.

Assessment of interests in policymaking was also taken up by authors such as Simon (1997[1957]) who pointed out that the range of interests brought different understandings and types of knowledge to the table, some of which were recognised by the 'authority' of planning and others were not. Some interests will be offered well-defined or routinised opportunities and be able to voice opinion and engage with planning. This typically includes statutorily inscribed opportunities to comment on draft plans, or to object to specific development proposals. The scope and structure of a planning process, in terms of the inputs received, can significantly affect the likelihood of representative values and information being integrated into decision-making. This means that the combination and implicit prioritisation of considerations that are supposed to reflect the public interest are often obscured.

Interests in planning may also be defined more narrowly in legalistic terms, as a set of parties who have some statutory right to participate in planning processes. The narrower view of (legitimate) interests in planning expressed above downplays the problematic and restricted nature of opportunities to engage for some members of society, Furthermore, some interests may not have a voice – these are sometimes labelled as 'passive' interests, and the environment or future generations are examples here. Wider conceptualisations of interests as anyone 'affected by a policy or development' embrace this issue and may be legitimately adopted here. This quite deliberately allows for an extension of the discussion of interests in the light of diversified attitudes and populations, with scope to consider whether planning processes and outcomes are adequately reflecting a diverse and changing society and prompts the question: How does such diversity present both a changing demand and a widened role for planners?

It may be that specific interests are formally acknowledged by planning law and policy at a local or national level, or there may be self-organised and critical action groups intent on influencing process and outcomes in terms of either preventing development, advocating it, or in shaping or reshaping policy proposals in some way and they may coincide with one or more interests. Such groups may be specially or unforeseeably affected by policies, plans and associated development or require active intervention by planners or other representatives.

The range of group and sectional interests have mushroomed, as diverse identities, beliefs and needs are recognised. Group-based interest claims are becoming not only more numerous but also more developed in terms of their sophistication, knowledge and tactics. Of course, some interests may not have access to the information, time or understandings necessary to engage effectively with planning. This latter concern underpinned the idea of planners as advocates whereby different interests were assisted or represented by planners in order to ensure fair access to planning processes. Indeed, this provides justification for planners to intervene in decision-making arenas that are dominated by the powerful (see Forester, 1989). Conversely, some minority interests have become adept at getting their views and position across at the local and national levels. As part of a political or economic (or even cultural) elite, certain sectional interests have been given privileged access and influence to decision-makers (see Cullingworth and Nadin, 2006: 446–7).

Interests, in this view, may be based on a variety of factors, for example locality, activity, or from an economic, environmental perspective, or they may be specific companies, groups or individuals involved or affected by development proposals directly. Equally the interest could represent a class or type of stakeholder in planning, for example the community, developers, landowners or even 'the wider interest', broadly following the categories outlined above, or even broken down further, given the now well-developed critiques of planning based on class, race, disability and gender inequalities. This highlights how, when we talk of 'interests' in planning, we may actually be referring to shared interests on behalf of others, in terms of a national interest, or a sectional local interest – in say housing or a local landscape, or indeed the self-interest of an individual.

How different interests affect or are affected by planning decisions will vary from place to place, as well as having influence at different times and in terms of different issues or characteristics. However, the term 'interest' can reflect the idea that an individual or group has something to lose or gain from a particular decision. Actors may seek to maintain a status quo or to strategically introduce alternatives. Therefore some organisations set up to represent an 'interest' may take a strategic stance towards planning policy and development with an 'interest position'. Typically large development companies will lobby for policy that creates opportunities for profitable development, while other groups will lobby for more restrictive polices, for example, the CPRE seeks to curtail development in rural areas. Both may claim to represent or fulfil the public interest in some way and seek the support of planners by mobilising a claim to represent a widely supported interest position. The contestation of the public interest reflects the sectional and political nature of planning, with claims and arguments about the benefits or disbenefits of particular policies and development proposals being associated to competing conceptualisations of the public interest.

There is a question, therefore, about how different groups or interests are able to act on opportunities or 'rights' to participate in planning and how some carve out other methods and means of influencing planning processes. The relative power and influence of different interests and their supporters will vary and the way that planning systems and structures are organised tends to benefit some over others. The question is: Does planning truly enable an articulation of all interests and therefore arrive at a justifiably labelled 'public interest' outcome? Some have criticised planners for shirking from a full discussion of the issues and alternatives, belying the professional and historic elitism of much decision-making. Thus when we consider the various underpinnings of the public interest, it is clear that different uses and approaches may be justified, or considered desirable, following different conceptualisations of the policy process and of institutional design – that is, depending on degrees of inclusivity and deliberative credentials (see Ham and Hill, 1993, and the wider literature on collaborative planning, for example Healey, 2005). We now turn to consider how the idea of the 'public interest' has been interpreted and used to justify planning decisions.

Planning and the Public Interest

The idea of the public interest has received considerable attention over the past decades (Benditt, 1973). There has been a resurgence in attention more recently in the UK in response to political attacks on planning during the 1980s and more recently in the context of increasing social diversity. These challenges and responses have cast doubt on the existence of a cohesive or general 'public interest'. As well as promoting defences for planning as part of a wider endeavour to serve the public interest, this has been fortified by claims that planning can help promote social and environmental justice, that is, that without some form of planning and intervention spatial (and socio-environmental) outcomes will be both unfair and inefficient.

Planning policies are often justified in terms of their overall benefit for the wider population. In this view, the outcomes of planning actions *should* be beneficial in terms of an aggregate social good or in some utilitarian sense. This role is seen widely as a key justification for planning systems and as part of this, it is assumed that planners will seek to minimise detrimental outcomes. However, the key question here is *how* is the public interest evinced and maintained through principled positions, institutional design and policy detailing?

The imperfect institutional design of planning systems, and an awareness of their limits, has meant that underlying assumptions and the rhetorical use of the public interest require some interrogation. There are numerous arguments and implicit conceptualisations that planners and others could adopt and do actively draw upon. The emphasis has tended to be on claiming that the planning system is 'fair', as well as being strategic in terms of achieving goals set across a territory. In this sense the public interest is equated with participation opportunities through a reliance on open and transparent *process*. However, there is clearly more to the public interest than ensuring 'fair play' and hoping that the outcomes are satisfactory or consistent.

Several key authors have recently examined public interest and have usefully attempted to unpack the concept in terms of the way that planning has both performed and justified itself in processual and substantive terms. Campbell and Marshall argue that planning needs: 'resolution of the ethical dilemmas which confront [planning] not just in procedural change, important though this is, but in redefining the public interest in terms of the purposes and values which the planning system is seeking to fulfil and promote' (2000: 309).

Alexander (2002a) suggests that the public interest criterion has three overlapping roles. Firstly, to legitimate state activity and policy, such that those actions are shielded from attack by individuals and sectional interests. Often this is done by recourse to a utilitarian defence, that is, that it serves the majority of the population or recognisable (and preferred) interests. Possibly, and if deemed necessary, government may invoke the 'national interest' to override local objections by recourse to the principle of subsidiarity; for example, where major infrastructure is planned. The Minister for Local Government, John Healey, in England made the following statement in 2007, which highlights this hierarchical approach and the assumption of the national interest in respect of major infrastructure:

> Decisions are required at the right level. Those which affect us as a nation should be taken at a national level, so that the wider national interest can be considered. Others should be taken at a regional or local level. And political decisions – such as infrastructure policy – should be taken by politicians. (2007: n.p.)

The second element is that the concept serves as a normative and foundational basis for practice whereby planners and professional institutes recognise and adhere to planning practices that seek to further the 'public interest'. The third aspect relates to using public interest as a measure of *outcomes* and to assess the substantive performance of planning. That is to say whether society is being well served or not in terms of the outcomes and benefits derived from planning practices. In essence this latter dimension asks: Does planning deliver a better society, better environment and a better economy? And is this done in line with what the population want? So in essence this becomes a wider question of 'what value is planning to society' as well as what values underpin the practice of planning.

There are several key dimensions and philosophical underpinnings implied here, each of which highlights how this phrase is used and can be understood in terms of the implications for the process and aims of planning actions. Alexander (2002a) distinguishes in some detail how the three roles or aspects of the public interest concept in planning can be assessed. He considers seven perspectives ranging from utilitarianism, where the action may be justified on the basis of whether it is 'beneficial', to a 'deontic' position where the key question centres on whether the planning action is morally 'right'. He also sets out a 'unitary' view of the public interest, claiming that this perspective is the one which underlies much current practice. This is where both a consensus is sought and a set of normative objectives are espoused.

Campbell and Marshall (2000) also see three conceptualisations of the public interest that broadly parallel Alexander's work. The first is a utilitarian view that avoids any moral judgement about individual preferences and acts in the interests of the greater good. In such an instance, this application is a rather crude summation of interests whereby the majority view prevails, or where a set process and local and national policy is applied to determine the correct course which will comply with the public interest 'test' (see Chapter 6). In recent decades the public interest has been equated with greater individual freedom in the UK, with the market mechanism used as the primary means to exercise that freedom. The second view prioritises the idea of public interest as being rooted in shared values where those values are applied to modifying market processes in the interests of the public good. It also makes space for the idea of acting on *behalf* of interests with less power or knowledge, either in terms of their own interest or of the (negative) impact of others' actions based on their self-interest. The third view places the need to ensure fairness of process as paramount. Such a procedural public interest has been the main concern of normative theory and the communicative or collaborative 'turn' in planning theory. This implies that the priority is to ensure that the planning process is inclusive and fair.

It is the often implicitly 'unitary' view of the public interest that Alexander identifies where the public interest is based on some shared or collective moral position; together with adherence to a predesigned and often rigid process for calculating the impacts of decisions. Despite planners and politicians adopting such positions, the relative weakness and enforcement of these against other powerful interests means that the intentions underpinning a planning system may start from this kind of (flexible) unitary position and claim to apply this; but often the decisions look and feel far more utilitarian. A question also is left open here about just how the moral or substantive goals of planning are defined in the first place.

In the UK efforts are made to promote procedural fairness and a form of unitary or shared moral and substantive purpose, but there are perceptible failures, gaps and weaknesses. Our view is that the public interest tends to be used more in a pragmatic and politically convenient or expedient way and any espousal of clear values or an inclusive process is elided rather to allow for 'discretion' and room for negotiation. The problematic use of the public interest as the justification for the actions of an elected politician is also uncomfortably evident in practice, that is, where the public interest is used rhetorically to support whatever the democratically elected representative prefers. This conflation of the politician representing and voicing the public interest becomes self-evidently flawed when they act as short-term vote maximisers, rather than adopting longer-term or responsible positions based on evidence, careful deliberation and in reference to overarching goals such as sustainability

Ideas of the public interest have also been attacked in the light of social and cultural difference and the perceived inadequacy of traditional policy processes and bureaucratic models to deal effectively with diversity. This is perhaps particularly true in terms of calculating the interests of disadvantaged groups and minority interests, or with other passive interests. Implicit assumptions of one definable single public interest can become a rather lazy excuse for predesigned action to be implemented, or leave way for an elitist conceptualisation of interests. This, at its worst, essentially absolves politicians and other decision-makers from needing to understand the preferences or needs of those whom they 'represent', or of fully understanding the principles underpinning planning.

Efforts to ensure that rationality and public choice are developed and understood better are important, if nothing else to reduce conflict and enhance legitimacy in planning. Redesigning planning systems that can develop adaptive or reflexive learning about needs, options and outcomes for a diverse set of interests is one future option. Meantime, one attempt to modify planning processes, to allow for such critiques, is to extend formal rights to participate in planning, and there is a wide-ranging literature on this. One aspect of this is the notion of extending 'third party' rights (see Ellis, 2002, 2004, and Chapter 12). Third party rights are entitlements to engage in formal processes of decision-making over development that sit beyond the developer and the planning authority (as the first and second parties) – the idea being that checks and balances can be introduced to ensure that due consideration to all interests has been given. One formulation would allow a third party the right to appeal against a development decision. Such rights have received renewed interest in the UK in recent years, but were resisted as of 2011. Instead the developer retains the right to lodge an appeal for a refusal of planning permission while other interests cannot do this, or conversely to appeal a granting of approval. Such prioritisation of one interest over others has been a longstanding area of contention and is justified on the basis that the planning authority itself considers and represents all other interests, that is, the unified 'public interest'.

Conclusion

The way that the public interest is said to be served is closely related to the stated expected outcomes of planning, as well as the processes followed in planning. A closer scrutiny of how developers, government and others set out the benefits and disbenefits of planning policy decisions and subsequent development patterns, is rarely achieved or attempted by planners. Cullingworth archly puts it that: 'planning proposals are generally presented to the public as a *fait accompli*' (1964: 273, cited in Cullingworth and Nadin, 2006: 355). Rarely are proposals the result of extensive and inclusive deliberations, or discussions which clearly demonstrate how the different interests will be affected. Opportunities for debate and engagement may be sub-optimal for pragmatic reasons, such as cost and time pressures, but also more worryingly perhaps due to the influence of powerful groups acting to exert pressure on planners and politicians at all levels. This influence has been most often applied within the scope of the law, although in such a way that those with power and resources claim understanding of the public interest and shape the agenda. The ability to orchestrate data, marshal (particular forms of) evidence and the ability to shout loudly are all likely to influence decision-makers' appreciation of what may be in the 'public interest' at the expense of less organised or visible interests.

Given this situation, an awareness of the different models and theoretical justifications for the public interest is necessary and the adoption of different processes and structures will mirror a preferred planning philosophy. There is a clear need to incorporate not only opportunities for 'fair play' in process terms, but also to firmly understand and agree the principles and aims of planning actions: as all good plans should. Opportunities to shape and engage with policy and development proposals will require ongoing thought and scrutiny, with deliberative approaches to public participation in planning meriting more serious and determined attention in helping to centre or 'ground' public interest.

FURTHER READING

Some key papers that dissect the idea of the public interest in planning are those by Alexander (2002a, b) and Campbell and Marshall (2000, 2002). Friedmann's (1973) paper is also still useful, as is Klosterman's work (1985). Booth (2002) examines how the public interest and planning control have developed through time, while a wider consideration of the public interest, and away from planning specifically, is found in the edited volume by Friedrich (1962) and in Lewin (1991). Related topics concerning ethics and values in planning can be discerned from a variety of sources, including Thomas and Healey (1991) and Marcuse (1976). Rydin (2003) devotes a whole chapter to lobbies and interests and discusses some of those typically found in planning, as does Bolan (1983). Healey et al. (1988) predicate their dissection of British planning practice on multiple dimensions of interest and this is picked by Adams (1994). Finally, Cullingworth and Nadin (2006) briefly refer to interests in Chapter 12 of their textbook on UK planning.

10 NEGOTIATION

Related terms: mediation; bargaining; conflict resolution; mutual adjustment; satisficing; partnership; policy process; consensus-building; collaborative planning; decision-making; communication; discussion

Introduction

Negotiation is a widespread feature of social life, with observers claiming that negotiation between individuals is essential to the operation of all societies (Gelfand and Brett, 2004). While many negotiative skills and strategies have been formally refined and studied or practised, it is the more unstructured or diffuse way that everyday practices of negotiation are commonly experienced. Negotiation may be seen as a practice and as a set of skills which are applicable in a wide variety of professional contexts too. Learned negotiation skills may be deployed with forethought, or be drawn on more organically to reach agreement, maximise benefits or develop shared understandings. Negotiation is commonly required to secure agreement on policy objectives, to achieve smoother resource allocation and more effective implementation outcomes in planning and development. As such negotiation processes play an important role in producing viable, more widely understood and more consensual outcomes.

Negotiation is a longstanding and common feature of the planning environment and merits inclusion here as planners will need to exercise negotiation and mediation skills in numerous instances. Cullingworth and Nadin assert that there is likely to be 'an even greater role for flexibility and discretion' (2006: 4) required in planning practice, and this logically requires forms of negotiation. Participants in the planning process need to recognise and understand where negotiation plays an important part in planning processes via brokerage between various interests (see Chapter 9) and where effective negotiation can help with policy implementation.

The politically charged and multi-stakeholder environment that planners inhabit means that very often negotiation can be complex and may appear to be rather thankless. In some planning situations where negotiations are brokered, important public, as well as private, stakes are being presented and refigured. In contrast to the view that 'the single most powerful tool for winning a negotiation is the ability to get up and walk away from the table without a deal' (Mackay, 1996: 81), it is often the case that planners cannot, practically speaking, walk away from negotiations. Rather, it is usually the developer that ultimately has the power to withdraw or disinvest from the process, with the planner's power lying more in an ability to delay. Forester cites an American planner thus: 'time is money for developers. Once the money is in, the clock is ticking. Here we have some influence' (1987: 304). On the part of the community, or other third party interests, there may not be opportunity to effectively delay or withdraw from a process, but such interests may instead have to live with the consequences of a failure to agree, or with the results of a poor outcome.

The extent and function of negotiation in planning has arguably changed, with a more negotiative and collaborative model of planning developing in some countries. Pressure for this has been felt particularly in the UK (Ennis, 1997), partly in the light of a more market-led planning that has emerged since the 1980s, although this dimension has been paralleled elsewhere too. A diverse and fluid societal context is another relevant shift and the collaborative and inherently negotiative approach to policymaking and urban and regional development, as expounded by Healey (2005) and others (e.g. Adams, 1994; Booher and Innes, 2002) encourages negotiation as a necessary part of a more equitable and democratic planning amongst diverse communities of interest.

The underlying issues and drivers of negotiation and the main justifications for negotiative practices in planning relate to a number of other chapter subjects here (e.g. Chapters 2 and 7). Ham and Hill claim that 'the study of the policy process is the study of conflicts between interests'(1993: 188). Given this situation, planning and associated policymaking and its implementation are centrally concerned with negotiation. Authors such as Claydon state that 'successful implementation is dependent upon successful negotiations among … interested parties'

(1996: 111), and this underlies attempts to integrate market demands and community preferences into policy and decision-making.

What is Negotiation?

Negotiation can be defined as a process whereby parties or groups attempt to resolve matters of dispute by holding discussions and coming to a mutually agreed decision (Fowler, 1990). Sometimes the *process* of negotiation can be as valuable as the outcome, however – in the sense that where agreement is not reached then at least further understanding between parties has been brokered. It may also be that through informal discussion it is discovered that the aims or needs of at least one party cannot be met and therefore time or expense, or both, may be saved (for example during 'pre-application discussions' in England: see Allmendinger, 2007; Beddoe and Chamberlin, 2003). This implies that negotiation can be useful in resolving disputes and achieving agreements, but also in communicating positions between competing objectives. This view of negotiation goes beyond the usual 'win/win' or 'win/lose' outcomes that feature in much of the generic literature on negotiation (see, for example, Fisher et al., 1991; Johnson, 1993; Maddux, 1999).

There is an important distinction to be drawn here between bilateral and multilateral negotiations. Bilateral negotiations can be prolonged, but where there are only two parties or interests the process can be relatively straightforward, involving linear or dialogic proposals and counter-proposals, with direct trade-offs being made. Multilateral negotiation involves more than two parties and usually features a complex meshwork of relationships circling around problems that can be difficult to resolve. This situation often requires a process of revisiting and checking over putative areas of agreement and shifts in position can alter others' negotiating stances in complex ways. Planning negotiations are often of the multilateral type, although there are also clear examples of bilateral negotiations between, for instance, planners and developers or planners in local and central government. In both types of situation there are several key skills required, as well as certain styles and/or strategies that can be used to fit particular situations. These are commonly cited in the negotiation literature, and include an ability to behave appropriately and to ensure that various preparatory steps are followed. Typically a four-stage model is used, following the: 'Prepare,

Discuss, Propose and Bargain' model (see Johnson, 1993), with different strategies and tactics being employed. A lack of training and skills in this area has become apparent in UK planning and undoubtedly in other countries. Regular training events and conferences are held to help tackle this perceived deficit. Making this point, a 2008 report into planning agreements brokered in London found that inadequate negotiating skills were causing a problem in ensuring that community benefits were secured (GLA, 2008). Without the necessary awareness and skills, the efficiency and adequate representation of the public interest by planners (Chapter 9) may also be called into question and equally others with less knowledge or power will undoubtedly lose out in what may otherwise appear to be a pluralist (and therefore apparently equitable) approach to decision-making.

There is a wider context provided by the communicative and collaborative turn in planning (see Forester, 1987, 1993; Healey, 2005; Innes and Booher, 2010; Sager, 1994), which focuses on power relations and interactions between parties and emphasises the process used to develop shared understandings in planning and development. We also seek to raise awareness of the importance, and provide examples of the practice, of negotiation in planning. This is considered useful to highlight negotiation both as a set of *skills* and as a *process*, and furthermore to understand the pervasive role that it plays in many planning systems.

While negotiation has always featured to some extent in planning, an emphasis on negotiation and on conflict mediation has accompanied the communicative turn since the late 1980s. Significant in this process has been the work of Forester (1987, 1993, 1999); Innes and Booher (1999, 2003, 2010); Elster (1998); Sager (1994) and, in the UK planning context, Healey (2005), who have all facilitated and reported on this shift. Healey suggests that, 'through ... a process of "learning how to collaborate", a richer understanding and awareness over local environmental conflicts can develop from which collective approaches to resolving conflicts may emerge' (2005: 34). This implies a collaborative model where different interests negotiate jointly in order to build some form of consensus.

Negotiation is a key feature in planning in many countries, although perhaps more visibly, in the planning systems where there is a considerable degree of discretion present (see Booth, 1996). However, there are many other opportunities to exercise negotiative skills in professional life and taken together, negotiation skills and the exercise of formal and informal mediation of interests form a key part of the planner's toolkit.

Many resources detail the approaches and tactics available to negotiators and we do not want to simply reproduce these. It is worth highlighting that those involved in planning need to understand basic behaviours and skills required, including negotiation strategies, methods and manoeuvres that will enable positive outcomes.

There are many opportunities to engage in negotiation during the formulation and implementation of planning policies. It is not surprising that professional organisations such as the RTPI identify negotiation skills as being central to effective planning practice. Some of the situations that planners negotiate within include.

- Conflict mediation and mitigation – to find areas where agreement or compromise is possible and to develop some understanding of opposing positions (such as in neighbour disputes or around new housing development, where *inter alia* questions of design, scale, layout, infrastructure or tenure are discussed).
- Resource allocation – to bargain to ensure that outcomes provide a share of resource or achieve objectives that are fair, equitable or the best allocation under the circumstances (with regard to planning agreements or bidding for government grants, for example).
- Advocacy for underrepresented or less powerful groups – to ensure that their voice is heard and their interests are treated fairly (e.g. during the provision of affordable housing, or when actively seeking the views of minority groups in plan preparation).

A range of aids to assist negotiation and mediation have been developed, primarily to visualise and engage with different groups. For example, the outputs from Geographical Information Systems (GIS) and design visualisations have begun to play an important role in identifying and mapping particular needs spatially to show the potential effects of development proposals. Techniques such as 'Planning for Real' and wider visioning exercises have been utilised to get people actively involved in the decision-making process (see, for example, Shipley, 2002). This more imaginative use of supplementary tools helps develop the understanding and necessary 'bargaining counters' that can be deployed to support a more inclusive negotiation process.

Some commentators have critiqued negotiative or collaborative planning as masking power relations, citing how such processes may co-opt rather than empower participants (Tewdwr-Jones and Allmendinger, 1998; Huxley and Yiftachel, 2000). Equally some parties may approach the bargaining process with established and articulated positions that are developed as defensive mechanisms, or deploy spoiling tactics. Much conflict over planning arises where fixed positions are adopted and discussions cannot take place to develop understanding and cooperation or such interaction is left very late in the day. Furthermore, if part of the planner's role is to be facilitatory and act in a mediating and developmental role, then enhanced communication skills and nuanced understandings will need to be brought forward.

It is important to also outline other related terms that form part of discussion of negotiation. The key separation that we wish to highlight here is between that of negotiation where the planner may explicitly, or otherwise, act as an interest with a stake in the outcome in his/her own right, as opposed to the planner as *mediator* who is acting, more or less, as a neutral party and is helping to orchestrate agreement between other parties. In *negotiating* a particular outcome the planner may typically claim to represent the public interest, or perhaps one preferred option in terms of a development proposal or a policy design. However, the planner may also be in a situation where they are asked to articulate the local planning authority's position, or the interests of the ruling political bloc, which could compromise the ethical and professional judgement of the planner involved (see, for example, Healey, 2005). As Forester indicates (1987) there may be confusion between these two notions; and, furthermore, there may be instances where planners are simultaneously trying to mediate and negotiate. This may give rise to a conflict. In the *mediating* role the main aim is to ensure that all parties can find voice and legitimately represent their interest, as with advocacy planning, while in the negotiation role the main aim is to realise the position of the interest represented.

Another distinction that can be drawn is that of the planner acting *between* interests *or* negotiating *as,* or on behalf of, an interest. Regardless of the positioning or interest being represented, the skills of negotiation are widely deployed in planning and are required by all interests involved. Given that attempts to manipulate, co-opt or to exploit other interests tend to persist in planning across the world, a consideration that requires our attention here is, firstly, the *representational* role of the planner, where the planner is acting on behalf of absent interests (often conflated under a rather more opaque notion of public interest as discussed in Chapter 9). Secondly, we are interested in both the *process* and the reasons or motives for negotiation in planning. Thirdly, we are interested in the *techniques* and benefits of negotiation and conflict resolution. Good negotiation should seek to achieve a consensual outcome, or at least an outcome where everyone involved understands *why* the consequent decisions were made. This leads us now to consider why interests negotiate in planning.

Why Negotiate in Planning and Development?

People tend to negotiate when they need or want something that requires the 'agreement' of others. Benjamin Franklin famously said that 'necessity never made a good bargain', highlighting how one's negotiating position and the relative power of the interests involved in the negotiation will almost certainly affect the outcome. We can discern that, in simple terms, interests negotiate when there is some power held or resource that can be traded or in some way combined. There are other reasons why interests increasingly seek to negotiate in a planning context and the need to ensure quality of outcome in a discretionary system is a key factor. The abandonment of state-led 'blueprint' planning approaches in many countries has meant that negotiation is more commonplace in order to aid decision-making in planning and/or to facilitate a more open process and equitable outcomes.

Satisficing is a term commonly used in the literature on the planning process and implementation, and it reflects a common outcome of negotiation and bargaining in planning, whereby compromise or 'mutual adjustment' (Lindblom, 1965) takes place. It is a portmanteau word joining 'satisfy' with 'sacrifice' or 'suffice', indicating how different interests may accept suboptimal outcomes as the best possible result in the prevailing circumstances (see also Simon, 1997); this reflects how a satisfactory compromise may be reached through a consensus-building or negotiative exercise. Ennis argues that: 'decision-making cannot be made in a rational fashion because what planners have to do is to take account of the differing and conflicting values held by individuals and groups in society and by planners themselves' (1997: 1943). This stems from a realisation that became undeniable by the 1970s; that the development industry were often leading the process of regeneration and of economic development, rather than the state, and that it could halt key public policy objectives more or less at will.

Given this context, negotiation becomes important to 'oil the wheels' of planning and serves to mitigate conflict over divergent goals.

Motives and Limits for Negotiation

The motives and impulse for interests to negotiate will be substantially shaped by the opportunities and practices afforded by the institutional environment. There will also be varying 'conditions of possibility' requiring knowledge and understanding, and which can lead interests to the negotiation table. Most typically this approach will be applied to a disputed or contested development proposal or draft policy, but informal or wider processes may be undertaken, particularly given the wide range of issues and contexts in which planners are involved (cf. Jansson et al., 2006).

Forester indicates that while there may be limitations and boundaries, including where legal or other equity considerations prevail, negotiation can be very important because, 'when diverse interests rather than fundamental rights are at stake, mediated-negotiation strategies for planners make good sense, politically, ethically, and practically' (1987: 312). Thus, in some cases there are some non-negotiable elements, as certain core principles will be untradeable, which will depend on stated or underlying ends that planners and wider society will wish to attain or defend. In many situations it is possible to arrive at outcomes that maintain matters of principle, or core aspects of a position, while negotiating around other elements – such as percentages of social housing to be provided, or the exact housing densities to be built, or numbers of car parking spaces for a new development, or indeed other restrictive conditions on use (see Duxbury, 2009). Public sector planners may therefore be acting in the wider (and perhaps more remote or abstract) 'public interest', as discussed in Chapter 9, but also and simultaneously for existing neighbouring communities. They may be representing absent parties such as the future users, tenants and residents of the new spaces that will be developed, as well as implicitly acting on behalf of the environment and wider environmental objectives developed through local, national and international policy or agreement.

Healey has noted that the model and context within which negotiation takes place will fundamentally affect the rules and the boundaries of acceptable process and outcomes (Healey, 2005: 224). The bargaining practices used also connect us to debates over exclusion from decision systems. Rarely do negotiations really allow time for mutual understandings to be developed. This hampers institutional awareness raising and consequential broadening of bounded rationalities (Parker, 2008; Simon, 1997). As Forester argues: '[negotiation strategies] are hardly 'neutral'. Planners who adopt them inevitably either perpetuate or challenge existing inequalities of information, expertise, political access, and opportunity' (1987: 312). Forester also made an important distinction between mediation and negotiation as outlined above, and he set out six mediation strategies that typically feature as part of negotiations in planning and in the context of land use conflict. These are idealised stances or roles that planners may adopt or play out (after Forester, 1987).

1 *Planner as a regulator* – where the role of the planner is limited and the main function is as information provider. The planner acts to signpost others and indicate what the rules are and what is desired or 'required' by the planning authority.
2 *Planner as an advocate* – negotiating on behalf of other interests. This is where the planner more actively represents and uses the concerns and preferences of other, predominantly community, interests when conducting a negotiation or forming a bargaining strategy.
3 *Planner as a resource* – whereby the planners act as facilitators encouraging other parties to meet and listen independently to each interest involved. The planner steers the negotiation towards some form of agreement. This type of role is less likely to take place in the UK or EU but is to be found in the USA.
4 *Planner as a diplomat* – acting as chair and operating to get the best deal for other parties – effectively acting as referee of the negotiation process.
5 *Planner as an interest* – negotiating and acting in the wider 'public interest'. This is where the planner is an active participant in the negotiation and may have their own or the planning authority's interest to represent. For example, to ensure that the targets and priorities – which are probably already established – are represented and bargained for.
6 *Planner as a team negotiator/mediator* – where planners divide between mediation and negotiation roles across and alongside one (or more) of the first five strategies.

The above strategies and behaviours demonstrate how planners can potentially perform in negotiative spaces. They also suggest how different circumstances and environments might lend themselves to different roles and dispositions. However, it leaves open the question of the constraints that are present and are acted upon by different actors by dint of knowledge and information, existing structures, rules and skills (Campbell and Floyd, 1996). Questions remain about the effectiveness and legitimacy, as well as costs involved, of planners performing some of the roles outlined above. What this means is that there will be arenas and situations in planning where negotiation is necessary and expected. It may be part of the formal structure or will have become otherwise routinised as part of established practices.

There may be other instances and occasions where informal negotiation is required or desirable, however, many of these negotiative encounters raise ethical questions about the role that planners play and whether they are compromising their professional 'independence' (Thomas and Healey, 1991). This becomes a question of the culture of planning, and moreover the maintenance of principles of equity and openness that many public policy processes aspire to.

Negotiation in Planning Practice

Negotiation skills are exercised in numerous planning contexts from interceding and mediating in neighbour disputes to acting to discuss strategic options with communities, developers and politicians. The range of these situations can shape the roles that planners play. The main example that we will examine here concerns the process of negotiating planning agreements in England, known in shorthand as 'planning agreements' (also referred to as 'section 106 agreements'). Such agreements involve more formalised negotiation over the details of development schemes and the range of planning obligations associated with planning permissions in England. These are enshrined in planning law and practice and are recognised as legitimate ways of ensuring that developments will be acceptable to all parties. This type of negotiation 'space' may take a lengthy period, involve several stages and be very complex, with a number of different parties having a stake in the outcome. Planning agreements and their history are also detailed in numerous texts and resources as detailed later.

The planning agreement example helps demonstrate just how simultaneously complex and structured negotiations in planning can be, as well as how important the discretionary approach is in ensuring that the uniqueness of developments and their context are considered. Planning agreements have been a feature of British planning since 1947, but until 1968 their use by local planning authorities was constrained by a statutory requirement to gain central government approval. Between 1968 and 1990 agreements grew in number in response to reductions in local government expenditure (especially for development infrastructure) and growing awareness of the externality effects of major development projects. Since then their use has been developed, refined and shaped by numerous test cases and iterations to law and policy (see, for example, Duxbury, 2009, for the legal scope and application of planning agreements). This refinement and normalisation was most notable after the passage of the Town and Country Planning Act 1990 (the full extant guidelines on the use of section 106 agreements and obligations were contained in ODPM (2005c) guidance at the time of writing, although it should be stressed that such guidance is subject to intermittent change).

Planning agreements are designed to ensure that developments are 'rounded' and as far as possible provide for the externality effects and needs arising from the impacts of that development. They will include both positive and negative obligations on developers. The idea is that the uniqueness and complexity of (larger) developments require discretion and negotiation to ensure that the best possible outcome is wrought. The rules and interpretations of the scope of negotiation are continually shifting with alterations to government policies in the UK and with legal decisions which act to reset the parameters. Healey et al. (1995) provided a thoughtful case study on how approaches to planning obligations and agreements differed from place to place. Such variance often made it confusing for different parties to navigate, and the process was opaque and lacked accountability. It showed not only how discretion in planning can produce uneven outcomes, but also that negotiation skills are important to assist in bringing about equitable outcomes, or sometimes just simply to broker workable agreements required in uncertain circumstances.

The typical negotiation of a planning agreement can take months to complete, requiring careful consideration and balancing of the priorities arising from local and national policy and politics, legality, cost, delay and financial viability. The types of things that may feature in agreements are: community infrastructure (such as community centres, schools, green space, affordable housing, restrictions on trading or operating hours, traffic movements, renewable energy provision, green travel plans and a host of other elements). 'Unilateral undertakings' also enable developers to propose such elements outwith the formal negotiation process; it is a bargaining counter only available to the development industry. This illustrates the importance of the 'rules of the game' in acting to structure the range and scope of negotiable elements and the power relations involved in negotiations. The key element in such negotiations is the awareness and knowledge of each other's positions and the scope for negotiation over, for example, percentages of affordable housing to be provided, or sums that could be paid for other purposes. This crucially requires knowledge and understanding of both policy, the legal precedents that exist and of the financial appraisal of development. Again, this latter aspect tends to give some negotiating edge to the development industry as in many instances this becomes the basis for negotiation.

Another example of negotiation is in the mediation of appeals through informal hearings in England (see, for example, Stubbs, 1997) and increasingly in the use of pre-application discussions prior to planning applications being submitted. More generally, such approaches are necessary when preparing and submitting bids for central government grants, working on regeneration projects, preparing design guidelines, enforcing planning conditions, agreeing management plans and remediating derelict or contaminated land. These provide spaces for negotiative planning and give local planning authorities opportunities to shape development proposals and improve the overall quality of places or diversity of activity (see Beddoe and Chamberlin, 2003; Carmona and Sieh, 2005). Such mechanisms can also help in clarifying the expectations and scope for development, which can speed up and 'smooth' the planning process. These situations require information exchange and buy-in from all the parties to ensure that a discretionary planning approach can work.

Discussion

Some concerns have been expressed over the extra time and costs that planning negotiations may place on different interests, particularly in discretionary systems and where opportunities to negotiate are built in or encouraged by governments or other relevant authorities. Booth

(1996), for example, points out that other, less discretionary, systems of planning may be more efficient in terms of time taken. However, in terms of outcome, it is far more difficult to discern whether the range of specific and general conditions and criteria that are addressed in negotiation are achieved in more codified planning regimes and whether they deliver better quality and more equitable outcomes. There is also a contrary argument to Booth's which points out that preparatory negotiations can actually save time and other costs in the long run, as objections and refusals can be avoided or mitigated using such negotiative practices.

Clarity over where, why and how both public and private sector planners are involved in negotiation situations is important, as exemplified by Forester's work (1982, 1987, 1993). As Ennis (1997) highlights, the conditions under which negotiations take place, as well as the policy environment that shapes this, are likely to have a significant bearing on negotiations. Bargaining theory is centrally concerned with how participants perceive and use power. As such, the rules and processes, as well as the mediation function of planners, in all forms of negotiations relating to planning process and outcomes, are very important. The concept of negotiation is important, therefore, both in terms of when planners are called upon to negotiate in the public interest and when they may be involved in mediation to ensure that a fair discursive environment is maintained so that resulting negotiations are as equitable as possible. In Habermasian terms, this means maintaining a framework or 'space' for negotiation. The planner's role can be particularly difficult when trying to represent and defend positions on principle and also when trying to strike bargains on other legitimate 'tradeables'. If skills and awareness of how and why one is negotiating are lacking, then the danger is that principles can be easily 'bought-off', sometimes without the planner even fully realising it. As this may often be done in the name of discretion and to enable economic development, guidance and procedural rules are necessary to ensure a degree of accountability to such practices.

FURTHER READING

We recommend that Forester's (1987) thoughtful paper is absorbed in terms of the principles and implications of negotiation in planning. Also, Healey et al. (1995) and Claydon (1996) provide a useful general overview of negotiation in planning in the UK and specifically in relation to development proposals. A number of papers have been published more specifically on negotiation of developer contributions and planning obligations (for example; Ennis, 1997; Farthing and Ashley, 2002; Healey et al., 1995) and these provide focused consideration of one context for negotiation. The breadth and diversity of other contexts in which planners may invoke or enable negotiation is illustrated in a useful discussion by Jansson et al. (2006). These authors set out an attempt to develop and apply a negotiation tool in environmental planning from an early and wider perspective. Stubbs (1997) discusses mediation and planning appeals, Booth (1996) makes reference to discretion and negotiation comparing the UK to other countries, while Allmendinger (2007) highlights pre-application discussion as a form of negotiative space around the development of telecommunications infrastructure.

It is worth also highlighting the numerous texts and online resources that more generally outline negotiation strategy, tactics and behaviours. Many are not, of course, written specifically for planning and development, but contain useful guides, insights and principles that typically apply: for example, see Fowler (1990), Fisher et al. (1991), Johnson (1993) and Maddux (1999).

11 MOBILITY AND ACCESSIBILITY

Related terms: movement, flows, globalisation, postmodernity, migration, cross-boundary, fluidity, flexibility, accessibility, social exclusion, networks, transport

Introduction

Recent changes affecting our understanding of mobility and accessibility have presented a series of challenges for planning. These changes have centred on the impacts of globalisation and the associated flows of people, information and capital; as political, economic and technological boundaries and barriers have been dismantled and refigured in significant ways since the late 1980s. Yet it is acknowledged that such shifts have a trajectory that extends much further back in time.

The relative mobility of people and goods has been transformed with significant implications for future land use and in terms of the relations between people and between them and places. Mobility in terms of mass and personal transport has restructured cities and has effectively acted to 'shrink space'. Journey times have been reduced and electronic means of communication and interaction are increasingly instant and borderless, bringing people 'together' through cyberspace. Such means of communication have also extended reach and interaction, and produced new sets of relations and flows, while disturbing existing ones. This has become a key concern for social scientists and particularly geographers (see Latham et al., 2009). As such, the scope of this chapter is not limited to questions of personal mobility but extends to consider wider flows and movements such as information and how these affect planning policy and decisions.

Some such considerations of mobility have been regarded as important since very early on in the history of town planning. By and large this has been seen in terms of responses to the physical movement of people and goods and the resultant need for transport and other infrastructure. Planning to respond to, or accommodate, flows of people and goods are therefore well understood priorities, given that such activities clearly have had profound impacts on local and national economies, land use and the wider environment. Traditional planning has struggled to cope as the world has further globalised and the pace of the above movements and flows has increased while regulatory barriers have been eroded. The ability to predict accurately or control these multiform flows and migrations and to plan for future needs also becomes more uncertain and complex.

Equally, reacting appropriately to existing flows of people, travel patterns and to changes in distributions of economic activity has proven to be a highly resource-intensive task, with traditional planning methods, such as forecasting, often leaving plans wanting or obsolescent, as well as subject to other political and moral questions. This also corresponds with a crisis in terms of wider planning assumptions, for example the basis for assuming a degree of homogeneity of populations and of consensus over priorities and social preferences, which has been eroded in the context of more mobile, dynamic, globalised, diverse and multicultural societies. This recognition of globalisation and fluidity calls into question not only the assumptions predominating in planning policy, but also the structures and methods of traditional planning practice.

A full account of the challenges and resulting reconsiderations of planning practices and theory development in this period cannot be rehearsed here; it is enough to restate that top-down solutions to many questions that concern planning are affected or destabilised by the repercussions of the global age. So far, most countries have found it difficult to reconcile 'planning' with a postmodern environment. In this light, and as if justification were needed further for planners to be concerned with mobility, Urry argues that the increasing role of trans-national networks and the rise in 'flows of people, money, and information' highlights that we should be studying mobility and its ramifications intently, and he has gone as far as claiming a new 'mobilities' paradigm in sociology (Sheller and Urry, 2006; Urry, 2000a; 2000b; 2002; 2007), given its far reaching significance for society in general.

This chapter examines how planning is influenced by, and responds to, new (and more longstanding) challenges of mobility. There has been a longstanding recognition of the need to ensure and manage accessibility to components of the built environment for different groups in society. So, while the lead concept is that of mobility, the associated idea of accessibility is also discussed to consider outcomes that a more mobile and globalised world may produce. For example, advances in communications and widespread car ownership have not only opened up new personal freedoms and experiences, but may have also left some social groups or individuals facing disadvantage or social and economic barriers. In terms of places, or the 'spatialities' of such processes, changing economic and political dynamics can equally marginalise or lead to disinvestment for some towns and cities (or conversely to the 'overheating' of other local economies).

Mobility and Accessibility: Definitions and Application in a Globalised World

Mobility may be applied to resources and flows of resources, as well as people and their physical ability to go from place to place. At its broadest, the idea of mobility implies the ability to move and to do so freely or without restriction. Questions of personal, social and economic mobility come into view as part of that very broad definition. However, there is a division to be highlighted between thinking about personal mobility effected on a routine or perhaps seasonal basis (and, for example, access to local services) and the grander notions of mobility concerning aggregate movements and flows of investment across political boundaries, or with people migrating between regions and countries.

Both mobility and accessibility imply movement or the (in)ability to use areas or resources. Accessibility is clearly linked to concerns with mobility, but it also infers the relative and changing ability to reach or be reached and is therefore connected to concerns with social and spatial justice. For example, efforts to ensure that services are accessible to different groups where some may have little or no access to transport or advanced communications technology (sometimes referred to as 'digital exclusion') may be issues that planning policy seeks to address. Indeed, some interests in protected areas may try to prevent the extension of transport or communications infrastructure in their areas in order to dampen development pressures and retain landscape quality. This throws up contradictions and challenges for planners in thinking about possible conflicts between social, economic and environmental sustainability (Chapter 3). The variables and impacts implied by the mobility of people and capital mean that mobility carries associated social, environmental and economic consequences; and this is why planning has traditionally been involved in managing such movements.

The organisation and orchestration of different flows and mobilities have an influence on wider socio-economic dynamics and may be shaped to some degree by planners. Moreover, the wider economic and spatial impact of mobile or 'footloose' capital has had a marked impact on the ability of strategic planning to direct, and on attitudes of many governments towards regulation. Part of this concerns the difficulties that planners have in keeping up with, and providing appropriate prescriptions for, quickly changing circumstances. This is often expressed in calls for the reorganisation of spatial governance of planning or by shaping planning policy around such forces.

Globalisation and its associated impacts on spatial relations has clearly changed patterns of land use and demand for development and opened up possibilities for new spaces and places. Thus, both of the central ideas discussed here involve questions of movement and change, and also confront some fundamental issues found with traditional methods of planning and plan-making (as well as the role of the state). Wider shifts in global governance and capital flows, for example, have acted to problematise traditional assumptions and practices of planners and area-based plans. The traditional assembly of plans for given bounded spaces are undermined when change is rapid and where flows and network relations extend beyond those boundaries with more unpredictable and complex possibilities.

The impacts and implications of enhanced personal mobility are linked, of course, to the *relative* mobility and accessibility that present and future physical environments and infrastructures provide for people (and which may influence such flows). As such this centrally involves how the (planned?) built environment attracts, retains and enables flows and mobility, and then affects local economies and places. A broader question here is: How to plan in conditions of enhanced mobility, 'footloose' capital and cultural diversity? This question has occupied planning theorists and has resulted in ongoing debates over the character and feasibility of 'postmodern' planning and planning techniques. Those debates consider how and where forms of 'postmodern planning' could adequately cope with change and diversity and with the implications of new and contingent global–local relations and flows, and where decision-making should most appropriately lie. Much of this debate has polarised into arguments about altering processes of planning, or questioning the ability of planning to achieve its core normative goals in the light of the 'threat' that global flows and the new mobilities present to sustainability.

Efforts to orchestrate and direct these flows and mobilities cut across a concern to direct economic activity and investment, and also require thought regarding socially equitable distributions and likely impacts on the environment. Varying degrees and shifts in mobility and migration deserve mention and indeed the idea of relative accessibility is of interest. Primary drivers are the availability or absence of restrictions on movement, perhaps legal, financial or knowledge-based constraints and, of course, limits on various elements of infrastructure (such as airports or major roads). Planners will be centrally involved in providing infrastructure and other physical elements such as roads, offices, industrial space and housing that service flows or enable mobilities. Equally, access to intermediaries such as computer networks or cars plays a part here as they all influence mobility and the relative accessibility of people and places.

In order to put mobility into perspective, we briefly reconsider globalisation, as the term is a regular counterpart to questions of mobility. Globalisation has been debated at length, and various features and processes are implicated. Technological advances and the rise of computer technology, as well as rapid transport and widespread car ownership in the advanced economies in Asia, Europe and North America are common features and have aided flows and mobilities across borders. Political change and the near hegemonic influence of market economics and neo-liberalism have had an impact on mobilities in terms of flows of investment and people and therefore on spatial relations and land use pressures. This process has been aided where governments have adopted a free trade policy, which has led to an opening up of markets and enabled new flows of capital across traditional borders. The EU itself has a prime aim of developing mobility between its member states, to enable freer circulation of goods, capital services and persons.

This process of opening borders and removing trade barriers is leading to the development of a single global economy that has been anticipated for some time. Appadurai (1990, 1996), for example, envisaged that the key and emerging features of globalisation would act to erode the ability of nation states to control their economy and maintain old continuities. This has precipitated a shift of analysis, including viewing the role and influence of networks. Castells highlights the importance of financial networks: 'the network society is a capitalist society. This brand of capitalism is different from its historical predecessors. It is global and it is structured around a network of financial flows' (1996: 471). This opening-up process also highlights the cross-boundary nature of flows and migrations and how the reach and possibilities for different places has been extended. This has led to an increased emphasis and policy concern over global competition, as well as the changing relationships between places and across space, calling into question traditional hierarchies of place and problematising old structures of governance and the contradictions of historic boundaries. Critics argue that the geopolitical conditions of the networked society (i.e. under conditions of global capitalism) can be exploitative and can enable corporations and multinational firms to garner ever more power and influence over national and local governments. Such conditions can also provoke bitter competition between cities and regions in different countries, as each seeks to promote itself and attract investment and other flows, such as firm relocation and tourist spend.

The global market clearly offers benefits and disbenefits, yet companies with international reach can make advances to domestic authorities regarding job creation. Conversely, there may be implied or actual threats to disinvest in particular areas in the light of many opportunities for capital to relocate to find cheaper (or in other ways superior) conditions of operation. This can aid a disembedding of such actors and encourage little or no loyalty, regard or feeling of responsibility for any workforce, place or country.

In planning terms, such a situation has helped create uncertainty and a context where few assumptions can be safely made. In such conditions, plans and strategies have a role in informing investors and other actors or interests to attract or to retain investment, but these may be overridden as new opportunities arise or when circumstances alter. This means that one of the continuing roles for strategic planners, both in the UK and in many other countries, is to make available and clarify the suitability of land and provide appropriate infrastructure for development. In relation to how such global shifts and broad advances in technology affect localities, we can consider how improvements in transport infrastructures (i.e. roads, rail, airports) and car ownership have impacted on locational decisions and patterns of travel, including workplace location and decisions about where to live. Thus, these have knock-on effects for travel-to-work patterns and affect business decisions and employment levels in any given area. Indeed the provision of enhanced infrastructure is seen as an important tool in maintaining competitiveness. The quote below from an Indian newspaper article underlines how places around the world are responding to globalisation and associated mobilities:

> World-class infrastructure is not only the key to a globally competitive economy but is also critical for improving productivity across all sectors. Inadequate and poor infrastructure is the foremost constraint in India's economic growth. (Pandit, 2009: n.p.)

The quote demonstrates a particular anticipation of growth and development. Two processes are in tension in this context of mobility in a globalising world and in a situation where neo-liberal political and economic models are dominant. Firstly, the tendency towards convergence and homogeneity in terms of policies and systems is apparent on the one hand. This tends to urge market-led planning models and, moreover, stress that the needs of business and the 'economy' be prioritised over other considerations. Secondly, on the other hand, a fragmentation of societies and recognition of the divergent interests of a range of social and ethnic groups problematises traditional plan-making. This raises the possibility of competing or incompatible priorities, or the elision of such issues in favour of economic growth and spatial policy being more reactive or 'on the hoof'.

Mobilities and Planning: Applications and Issues in Practice

Planners have been involved in shaping, or their thinking is otherwise affected, by the results of mobility at different scales. We have highlighted that different flows and the new mobilities of information, people and products have radically reshaped socio-economic patterns, giving rise to consequential spatial and environmental challenges. In this context we now consider two examples, showing how planners act to shape mobilities and respond to the challenges of globalisation and cross-boundary flows. Firstly, attempts to overcome the limitations of bounded or narrowly focused plans and traditional planning methodologies are discussed through an explanation of city region planning in England. This is followed by a short consideration of transport planning, and the responses to pressures for enhanced mobility and reduced travel times, by looking at sustainable local transport plans (LTPs) in this wider context.

City regions and bounded plans for a fluid world

Early planning methods tended to infer a degree of socio-economic and spatial stability, predictability or continuity, and relied on positivistic population and demographic forecasting. Enhanced mobility in terms of commuting and a propensity for people to move around more frequently have meant that effective planning has become even more difficult, that is, the standard questions of what should we build and where become more problematic as in-migration, out-migration and internal migration takes place, and flows of investment follow contours of profit or risk minimisation globally. The organisation of planning regimes nationally and at local and regional scales has also struggled to deal with cross-boundary issues, and this is compounded as flows and functional relationships become less and less likely to correspond with old borders. Understanding the flows and areas where investment is likely or needed is an important facet of strategic planning and has occupied academics and policymakers for decades, but there is less understanding developed in many regions, or on the part of some national governments, about how to respond to and shape flows appropriately. The tools to direct these become all important and include incentivising and promoting particular areas to aid growth, or acting to ease pressure on other areas. The role of institutional design and governance arrangements to facilitate or adapt to such conditions also becomes important (Simmonds and Hack, 2000).

Thus individual municipalities or planning authorities are being encouraged to work across boundaries to bring together and organise 'functional economic areas' where the majority of people in a given region live and work, or which have the infrastructure to cope with urban growth. In many cases these same areas are also trying to direct inward investment and cope with the demands of international competition. City regions reflect an attempt to think in terms of functional economic zones; and have regard to global influences and the opportunities and challenges in terms of their sub-regional economies and the flows which affect them. The label has a longer, and indeed worldwide, history but only more recently has it been introduced formally as part of strategic planning efforts in the UK.

The city regions in England are semi-formal partnership arrangements set up to coordinate the different actors involved in the shaping and delivery of strategic policy objectives. They try to orchestrate matters such as transport and urban regeneration following agreed strategic priorities, particularly in economic and spatial terms. In England, city regions have been somewhat flatly described as:

> the areas around major cities. These areas form naturally [*sic*], as a result of patterns of travel to work, shopping, or leisure activity. They do not necessarily coincide with administrative boundaries, so may cover all or parts of several local authority areas. (IDeA, 2010: n.p.)

This description may explain the *extent* of a city region and while they do cross some local boundaries, such areas clearly do not 'form naturally', rather Scott et al. argue that city regions *as units of planning* are being organised and shaped for a range of reasons, such that:

> this idea can be observed in the forms of consolidation that are beginning to occur as adjacent units of local political organization (provinces, *Länder*, counties, metropolitan areas, municipalities, *départements*, and so on) search for region-wide coalitions as a means of dealing with the threats and the opportunities of globalization. (2001: 11)

The city region idea also reflects an awareness of the limits of planning at the national level and with planning for small areas in isolation. This approach towards cross-boundary planning is an example of conceptualising places from a systems perspective and trying to recognise, and plan for, their network connections. As an example, the Leeds City Region entity seeks to coordinate strategic functions and claims to reflect:

> the real economy [*sic*] for these districts: the boundaries in which businesses operate, supply chains function and communities live their daily lives. It is the area in which people travel to work, school, and leisure, and as a result, it has cohesive labour and housing markets. (Leeds City-Region, 2010: n.p.)

The ability to actually do this and to look outwards and engage with international flows and connections is somewhat questionable however (see Scott, 2001; While et al., 2004). Despite such critiques, the Leeds City Region is an attempt to grapple with new mobilities and global flows and was, itself, nested into a regional approach to economic development termed the 'Northern Way'. The Northern Way grouping justified its existence, and that of the city regions approach generally, in this way:

> by working across administrative boundaries, they therefore provide a clear basis for delivering better economic policy outcomes in areas such as transport, housing, skills, employment and regeneration. (2010: n.p.)

These strategic groupings are attempting to find ways of serving their own populations by looking upwards and outwards, as well as focusing on developing their own capacities and infrastructures in a 'joined-up' way. Such partnerships are still largely focused on domestic strategic planning and tend to rely on traditional methods of planning and coordinated marketing for their sub-regions. This is typically coupled with incentives and planning policies orchestrated through regional or national agencies, including the preparation of documents such as 'city-region development plans' (see Price Waterhouse Coopers, 2007). Part of the efforts at a city region or sub-regional level includes transport provision and we now examine one aspect of this in terms of mobility and accessibility.

Mobility and sustainable local transport plans

Planners have long been concerned with physical mobility in terms of transport networks and the planning of related infrastructure (i.e. roads, railways, ports and airports). This is tied to questions of anticipating and managing the levels and speed of movement and flows of people in and across territories. Understandably, concerns to provide transport infrastructure and shape different types of transport use are seen as important considerations when negotiating or orchestrating sustainable development, particularly in the period since access to personal transport has been widespread. In the 1950s and 1960s the response of UK planners was to design cities and infrastructure to accommodate and prioritise cars.

The impacts of changes in demographic distributions and forms of migration, coupled with the challenge of having large numbers of people living and working quite far apart, is an ongoing issue for many developed countries. This transpires to create overlapping objectives both in terms of strategic economic development policies, as mentioned above, and in terms of transport planning policy (Banister, 2002; Banister et al., 2000). Transport planners are now attempting to balance the sometimes competing objectives. These include maximising mobility on the basis of ensuring personal freedom and enjoyment of travel, the maintenance of effective and equitable public transport services and the creation of 'competitive' infrastructure (including airports and high-speed rail). They are also concerned to minimise environmental impacts in terms of carbon emissions and congestion by reducing trip lengths and encouraging 'modal shift', that is, to get people away from cars and planes and to use public transport; overall, to assist in achieving more sustainable urban and regional forms.

We look more specifically at LTPs here to demonstrate how public sector planners at different scales are trying to reconcile growth and the demands for new development with sustainability. LTPs have been around in England since 1998 and were made mandatory in 2000. The thinking behind LTPs is to produce strategic approaches to sustainable transport by working with the range of relevant actors, including the community, and to consider the different forms of transport required by the area. The UK government argues that such efforts are an important part of regional economic performance and the organisation of services for local populations: 'the Local Transport Plan is a vital tool to help each local authority work with its stakeholders to strengthen its place-shaping role and its delivery of services to the community' (Department for Transport, 2009: 5).

LTPs were supposed to reflect regional and national level policy and help organise the allocation of central government resources to achieve policy objectives formulated at those levels. An example the LTP for the sub-region around Leeds (in West Yorkshire, England) explains the role and aspirations of efforts to organise transport infrastructure and services, with the economic dimension used as a primary justification, as follows:

> the transport strategy must seek to make best use of existing infrastructure as well as developing the use of alternatives to the car in order to manage traffic growth and congestion and provide the connectivity necessary for economic competitiveness. (West Yorkshire Local Transport Partnership, 2006: 11)

and

> West Yorkshire deserves a transport system that meets the needs of local people. Ignoring congestion is not an option if West Yorkshire is to stay competitive and see jobs and housing grow. The Partnership recognises that we need to be ready for this situation and need to investigate and plan what measures would be required in the future. (West Yorkshire Local Transport Partnership, 2006: 79)

In these ways, LTPs (like many other types of plan) start to accumulate multiple (often conflicting) priorities and aims. Moving people around efficiently should provide environmental and social benefits and contribute to the aggregate competitiveness of a locality set within its regional context. As discussed above, how to do this is not always an easy task given resource constraints, information lags and the rapid and fluid mobilities already discussed. However, they are part of a loosely coupled set of plans and strategies that, even with the tensions and contradictions that these multiple objectives (and the complexities involved) may create, aim to tackle the implications, priorities and needs implied by the new mobilities.

Conclusion

The main conclusions to be drawn here are that mobile societies and mobile capital flows, aided by advances in communications, transport and associated technologies are providing various challenges for planners. These challenges can be split into at least two parts. The first is the challenge of anticipating or responding to transport needs and other implications of a more mobile society. For example, tensions exist between transport infrastructure and the environment, given the ability of large numbers of people to live and work (and play) quite distantly. Efforts to understand and predict such flows and mobilities are constrained and the ability of national governments to control or shape them sustainably is also somewhat limited. The second aspect concerns the most appropriate spatial level to undertake strategic and economic planning, where efforts to understand and shape flows of capital and other resources are increasingly global and cross existing administrative and political boundaries. In turn, these aspects challenge planning theory and practice to construct new ideas and techniques to respond effectively to these 'postmodern' or 'glocal' planning processes.

Thus, the concept of mobility encourages planners to respond to and shape flows of people and capital to provide efficient and desirable spatial outcomes. At present this is done through a contingent mix of regulation, partnership, incentives and reliance on market forces. Conversely, we recognise how such challenges are forcing a reflection on the institutional arrangements and epistemological basis of planning that reveals the limits and difficulties, as well as the importance, of seeking to shape and encourage more sustainable spatial outcomes in a 'glocalising' world.

FURTHER READING

In order to think about globalisation and its effects in this context, Bauman (1998) provides an interesting and accessible read. Questions of space and place are mentioned in Chapter 13 here but for a broader and more global take, see Latham et al. (2009). In terms of planning and mobility more specifically, the edited text produced by the Organisation for Economic Co-operation and Development (OECD) (Koresawa and Konvitz, 2001) provides a reflective consideration of spatial planning in a global age. Beyond these sources, planning theory texts have examined the role and design of planning in a postmodern age, for example: Allmendinger (2001, 2009) and Tewdwr-Jones and Allmendinger (2002). A good but high-level introductory read on city regions is found in Scott (2001) and also Hall and Pain (2006) on mega-city regions. For further material about Leeds and other city region examples, visit http://www.cityregion.org/index.html, while the Northern Way strategy and partnership is available at http://www.thenorthernway.co.uk/. In terms of LTPs, the Department for Transport's (2009) guidance on LTPs in England describes the aims and typical content of such plans. The example of West Yorkshire LTP used above is one source (West Yorkshire Local Transport Partnership, 2006), while Shepherd et al. (2006) and Bickerstaff and Walker (2005) discuss LTPs and provide a critique of efforts to integrate and legitimise LTPs.

12 RIGHTS AND PROPERTY RIGHTS

Related terms: planning systems; interests; society; private property rights; human rights; participation; conflict; negotiation

Introduction

Explanations of the concept of rights are often found in discussions about the development of the modern state. Wide-ranging considerations and analyses of rights are also made in legal studies and in the political science literature, where the development of rights and responsibilities as codifications of conduct are traced. Viewpoints range from rights being fundamental to societies and their organisation to rights allocations being oppressive and perpetuating inequality. Rights feature in numerous debates in anthropology and sociology about their significance in different cultures, given that they are social constructs open to contestation and debate. Rights have attracted attention from geographers as they also have a spatial implication, for example in terms of how movement or activities are regulated in different spaces and places (Blomley, 1994). In application to daily life, rights are routinely cited and claimed as well as debated and contested in the courts; where individuals, groups or institutions will seek clarification about what entitlements, and indeed obligations, stem from a claimed right in a given context.

Such a wide range and depth of interest in rights reflects the importance of this concept in societies generally and, as we explain, in terms of the operation of planning. Decisions over land use and development play a role in affecting rights distributions and the economic values and wider distributional impacts that are derived by rights holders and others. In particular this is discernible in discussing property rights and the indirect rights of others affected by the implementation of planning policy. We argue that the role and operation of property rights are central to reflecting on planning and conflict over planning decisions (see, for example, Bromley, 1991; Sorensen, 2010).

The concept of rights is explained in this chapter through a brief discussion about the basis, types and the development of rights. Examples of rights and their role and application to planning are provided to demonstrate applicability. Further reading on rights theory, and in terms of land use planning, can be reviewed in numerous supplementary sources such as Freeden (1991); Cooper (1998); Bromley (1991); Ellis (2004); Geisler and Daneker (2000); Pennington (2000); Rodgers (2009) and Sorensen (2010). The chapter starts with some definitional clarifications.

Definitions of Rights

In general terms, a right is a legal or moral entitlement to do or to constrain others in some way. In this sense there are positive and negative rights that provide entitlements and place obligations on others. Rights act to provide an important plank in the regulation of society and legal rights are those allocations which have authoritative status. Coleman (1990) places stress on rights as social entities and only existing where there is a degree of consensus over where the right lies and why it is justifiable in social terms. Legal and informal rights (or entitlements) are maintained by society for the purposes of resource allocation, safeguarding of order and to provide clarity – particularly in terms of the legality of actions. Legally binding rights and other customary or informal rights need to be recognised in order to be regarded as a right or from which to base a 'right claim'. According to Bromley 'a right is the capacity to call upon the collective to stand behind one's claim' (1991: 15).

T. H. Marshall (see Cranston, 1973; Marshall and Bottomore, 1992) in his seminal work on citizenship identified different types of rights: civil, political and social. These categories have developed, and rights have been extended, as the modern state has emerged. Rights were formulated over time in a complex and evolving (or emergent) social contract. This has been considered by numerous authors, with Giddens (1984), for example, adding economic rights to this list. Stevens (2007) explains that some rights are potentially open to all and enjoyed by all and the notion of 'universal human rights' (e.g. right to life, liberty, privacy or, arguably, some social rights, such as a right to education) are often cited. Civil rights are a wide subset of such human rights. The right to a fair trial is an example and the idea that someone should be able to challenge a decision made against them if something about the process is flawed is regularly promulgated. As such, this has been cited in reference to planning appeals and third party rights highlighted below. Political rights (and rights claims) include the right to vote and the right to free expression, while social rights typically include, for example, rights to healthcare, welfare benefits or to a clean environment.

It is often the modern state which acts as guarantor for individuals in allocating and establishing rules and structures for civil society to operate within. Contestation over rights and obligations is commonplace and consequently rights are rarely stable as societal preferences or tolerances shift and the curtailment and expansion of entitlements reflect such social and cultural change. Rights are enforced by the judiciary and other authoritative powers and sometimes by communities themselves by way of moral/cultural regulation of behaviour (see Bourdieu and Passeron, 1977; Cooper, 1998). Such judgements and social regulation also serve to shape and reshape rights over time and from place to place. Bourdieu provides interesting thoughts on how such 'rules of the game' are reproduced and challenged (see, for example, Jenkins, 1992).

Rights that are observed beyond or outside of the formal national reification of entitlements and obligations may be expressed as localised 'rights' or interpretations of rights (and this hints at why local planning authorities interpret and apply policy variably). Equally there are rights that are held to be universal and beyond the nation state to decide or disavow (e.g. 'natural' or moral rights which may or may not be enshrined in national constitutions, or embedded as tenets of national or international law). Human rights, as with other rights, are theoretically based in some formulation of shared values and moral justice. They are therefore representations of moral and ethical positions, as well as reflecting historical crystallisations of wisdom and experience, and crucially of relative power distributions. In this sense, it is clear that the existence and distribution of rights reflect the contours of societies, or of the preferences of dominant groups in society.

If rights serve as rules of interaction between people, they place constraints and obligations upon the actions of individuals or groups, only some of which have been cast clearly in law. The maintenance of particular allocations may fundamentally affect the distribution of wealth in society. For example, rights to inheritance may mean that capital is passed from one individual to another by dint of birth (or will), and this can perpetuate existing inequalities. It is not surprising, therefore, that there is often a difference in terms of the understanding or acceptance of rights distributions among different social groups.

Wider contextual changes may mean that the definition and exercise of a right will require review and alteration. This could be due to local cultural shifts, policy change, or reflecting legal judgements applicable on a national scale. For example the emergence of the climate change agenda as a key global issue has brought new pressure to bear on the boundaries, scope and privileges attendant on private property rights. Planners are often involved on at least two levels here in deciding the utility and entitlements that actually flow to rights holders: firstly, on a case-by-case basis when granting planning permission and, secondly, at the strategic level in influencing the future utility (and therefore impact on the 'value' of certain rights) spatially by zoning or demarcating land for new development or for protection. In this way, planners become an important intermediary between rights holders and the wider society.

Property Rights and Planning

Private property rights are a form of civil rights protected by the law and are found embedded in a whole range of assumptions about resource use and exclusivity and the entitlements that are linked to ownership. Property rights are often cited as important building blocks of modern societies (and for capitalist models of economic exchange); for good or for ill (see North, 1990). Property rights 'guarantee' that the benefits and utility of particular resources will be realised by owners, with numerous constituent rights or features seen as comprising 'full liberal ownership' of land. Honoré (1961) lists 11 such rights or 'incidents' of property including: the right to use, the right to the income and the capital derived from the land. Many accounts of property rights and associated entitlements emphasise how they change over time and are curtailed to balance private and public benefit. These can be viewed as expressions of the (changing) social relationships between people and the contingency of rights (see Geisler and Daneker, 2000).

There is a wider understanding about the benefits of property rights, for example: 'property rights do not exist for the sake of those people with substantial property holdings ... rights exist to serve social purposes reaching far beyond those who actually exercise those rights' (Sowell, 1999: 164). Our reading is that property rights become a mutual convenience, but that some social groups or interests may benefit individually more than others from the distribution and management of property rights. Planning powers and decisions affect the benefits that may accrue from ownership (and associated rights), and planners act as a mediator between competing rights claims. Hodge explains that the planning system in the UK is: 'a major factor in determining the rights enjoyed by landowners. Planning defines property. Changing social values and attitudes lead to changing rules of land ownership, often implemented through alterations to the planning system' (1999: 101).

A central role for planners lies in examining and understanding the externalities that may follow if certain rights are or are not curtailed (see Chapter 16). This plays out largely over the regulation of property rights and their maintenance and exchange, or where private claims to rights are made in contradistinction to the 'rights' of others. This reverse angle concerns the wider civil or human rights relating to property that seek to secure privacy and security of ownership (Allen, 2005; Denman, 1978; Massey and Catalano, 1978), but also the notion that there may be a superior interest to be maintained on behalf of society (see Geisler and Daneker, 2000 and Chapter 9). The role of planning in shaping and allocating benefits to property rights holders therefore puts the planner in the midst of a highly politicised and often high stakes environment with contested social, cultural, environmental and economic consequences.

In underscoring the impact of change on property rights Rodgers argues that, 'whenever legislation alters the allocation of utility rights over land, then a form of property transfer has occurred' (2009: 135). So, for example, whenever a planning permission is granted, a change in the allocation of benefit will flow from this. 'Betterment' is an example where the economic value of property rights is boosted due to the extension of the lawful use or utility of a site (see Booth, 1996; Cullingworth and Nadin, 2006 and Chapter 7 here).

The practical ability to exercise rights can be influenced by one's particular situation or changing economic conditions, or for other reasons such as a lack of knowledge or of other resources. There may also be attempts to obstruct the exercise of rights. An early passage from *The Hitchhiker's Guide to the Galaxy* is illustrative of this point, and is particularly pertinent as it is planning and bureaucratic 'interference' in private (property) rights that are held up for ridicule. In this scenario the house of the main character, Arthur Dent is to be demolished to make way for a new road. The relevant documentation has been displayed for public consultation, but it is claimed by Dent to have been kept in the cellar of the local government office: 'in the bottom of a locked filing cabinet stuck in a disused lavatory with a sign on the door saying "Beware of the Leopard"' (Adams, 1995: 20). The indignant planning officer on the scene is clearly surprised that this opportunity, this civil right, to comment and object to the new road has not been taken up, and in Arthur's subsequent resentment at the violation of his (property) 'rights'. This humorous example highlights how rights should be both enforceable and *exercisable*, and this connects us to questions of participation, process and equity in planning (also Chapters 2 and 9).

Plant indicates how property rights make for a good example where the rights or liberties of others are co-dependent, and that the allocation and entitlements that are legitimised by dint of property ownership affects others:

> Taking property rights as given in our society in which there are virtually no unowned resources restricts freedom of non-property owners to exercise their liberty. Hence the real question is not about the infringement of liberty. The question is rather whether, for example, the right to the means of life has priority over the unfettered right to property. (1996: 186)

This highlights how rights may ultimately depend on the ability to successfully defend them. Rights afforded to property 'owners' provide protection and imply a duty on others to observe the rights of those property owners. However, those property rights also form part of a wider social contract which implies some reciprocal responsibility. The operation of land use planning is a clear example where the state intercedes through, and directs, property rights allocations and their exercise; ostensibly in the public interest and to ensure that deemed important public goods are maintained. Planning decisions are, if nothing else then, calculations of benefit and harm about the exercise of private property rights and their impact on society in terms of the impacts in economic, environmental and social terms (see Chapters 3 and 16). As such, the role and consideration of rights and property rights in particular are important for planners and policymakers in terms of creating sustainable environments.

As mediator and regulator of some alterations to property rights and social or 'amenity' rights, planners are often caught in a rather uncomfortable position, that is, between competing rights claims which may be supported by a range of arguments, evidence and other considerations. This politicised space acts to shape the inherent value judgements being made, and planning systems and planners act as intermediaries in the exchange or modification of rights of different types. Plans also provoke challenges due to the policies and implications for the economic value of land and other rights (e.g. right to a healthy environment); and each land use and development has a different economic value or implies a different 'benefit stream'. The most obvious example of rights being modified under planning legislation was in the case of development rights that were effectively nationalised in the UK under the Town and Country Planning Act 1947 (see Booth, 1996; Cullingworth and Nadin, 2006). Cullingworth and Nadin explain the result of the 1947 Act in this regard: 'Development rights in land and the associated development values were nationalised. All the owners were thus placed in the position of owning only the existing use rights and values in their land' (2006: 23).This effectively put development 'under licence' and put power in the hands of state planners and politicians to decide which sites, which land, could be used for different purposes and the ability to limit or place conditions governing the extent of the entitlement in terms of the type or extent of the use (e.g. the range of legal uses, the size of new buildings, or for example, the amount of quarrying permitted or the siting of wind turbines on open land).

This shift of rights allocations was justified in utilitarian terms to serve the public interest, so that the overall orchestration of development could be organised by planners through formal plans. Other constraints placed on individuals were in terms of building design and variables such as height or relationship to other buildings. Furthermore, other environmental and social impacts would be factored into decisions. In principle, this legislation shifted the right to develop land and property from private control to the state, with local planning authorities acting as agents granting or 'returning' rights to owners to develop through the system of planning permissions.

Rights and compulsory purchase

Development rights are defined as those rights that allow landowners to change use or to proceed with new development over their own land; these may be restricted as discussed. The typical rights of private property were listed by Honoré (1961), one of which is the right to sell. Controversially, public authorities may, in many countries, compel private owners to sell the land or property as required. This is either known as compulsory purchase or the exercise of 'eminent domain' powers (see Azuela and Herrera, 2007; Jacobs, 1961). The state may act to vest land for public purposes when it is deemed necessary to develop or enable development to proceed. In England, there are numerous conditions and circumstances in which this is allowed (see Cypher and Forgey, 2003; Duxbury, 2009; Moore, 2010) with (changing) rules and legal precedents which define and constrain the use of such procedures. Such conditions and 'tests' have developed over time and vary in different countries. In England the Planning and Compulsory Purchase Act 2004 (part 9) altered, indeed widened, the grounds for using compulsory purchase orders (CPOs), stating that planning authorities will be, 'able to acquire land if they think the carrying out of development, re-development or improvement is likely to be of economic, social or environmental benefit to their area' (HMSO, 2004: part 9, para. 128). This is the broad justification for using compulsory purchase and is hedged by certain tests that were set out in Planning Circular 01/05 and which are intermittently revised. Such guidance helps actors to determine the legitimacy of a CPO (see ODPM, 2005b). This example brings into view the conflict that can arise between private interests and their attendant civil/human rights claims and consideration of the wider public interest which may claim a justification for a taking of private rights. There are examples where efforts to take land in this way have met with protest and in some circumstances there may be competing rights invoked in order to trump compulsory purchase/eminent domain claims. In Japan, for example, the national constitution hinders governments in acquiring land compulsorily (see Parker and Amati, 2009; Sorensen, 2010) and there are numerous other institutional, political and legal differences that affect the way that compulsory purchase is used internationally (Adams et al., 2005; Allen, 2005; Cypher and Forgey, 2003).

Planning and rights to participate

Rights that enable individuals to contest decisions in the planning system, or rights that relate to participation in the planning system and rights to engage and influence planning policies are usefully labelled as procedural or 'system' entitlements. Such participative rights (and wider rights to protest) provide important mechanisms to check and balance representative and expert systems. For example, 'rights to voice' in procedural terminology are where citizens may engage with, and object to, decisions and proposals as part of the planning system. The right to peaceful protest is a jealously guarded civil right in Western societies, largely as it acts as a counterweight to perceived injustices or inadequacies of formal processes, or to systems of representative democracy. Protest also acts to challenge boundary rules that may be inappropriate or obsolete. In planning terms, protest may force a reconsideration of the methods, processes, scope of evidence and range of debates that shape planning decisions (Owens, 2004; Parker, 2002).

The notion of 'third party' rights in planning relates to how different individuals, *qua* interests, might enjoy the right to make objections concerning a planning proposal. In UK planning, only those directly affected by planning decisions are permitted to formally object. Developers are viewed as the first party in a planning proposal with the planning authority as the second party, and the latter is supposed to act on behalf of the state and in the public interest in forming policies or decisions on development proposals. When a planning decision has been made, no one bar the first party may appeal the decision (except in exceptional circumstances, i.e. where judicial review can be won), and only rejected development proposals can be appealed. However, third party rights allow for challenges to proposals and decisions to be made by others and, if certain conditions are met, to appeal against the granting of permission. This represents a broadening of the boundary rules operating in planning processes. The justification used in the UK for a restriction of rights to object in planning are often pragmatic ones, typically related to time and resource limitations. Ellis (2004) argues how third party rights could act as an ethically and equitably balanced method of ensuring that planning decisions are fair and thorough, arguing that it is a matter of principle that a fair and democratic system can be challenged. Such third party rights could ensure that the range of interests locally and beyond can hold planners and developers to account for their actions (Webster and Lai, 2003). While there are concerns expressed about how such a system would work, and the impacts it could have on the speed of decision-making (and, therefore, on issues of economic competitiveness mentioned in Chapter 17), in the Republic of Ireland third party rights are already incorporated as part of the planning appeals system. This is mediated by a board of appeal (*An Bord Pleanala*) which considers the grounds for objections that are received and acts as arbitrator. The system has worked well, with fewer appeals being made than at first feared, and the approach apparently acting to improve the care and attention of developers and state planners in negotiating and preparing development proposals (Ellis, 2002).

The Human Rights Act 1998

Human rights are universal claims to entitlements and are explicitly set out at different levels internationally. At the European level, perhaps the most direct and relevant to planning, is the European Convention on Human Rights (ECHR) and the UK's formal adoption of this convention through the Human Rights Act 1998 (HRA) as implemented in 2001 (see Allen, 2005; Cullingworth and Nadin, 2006; Parker, 2001). The implementation of the HRA brought widespread speculation about how such entitlements could clash with existing planning practice and procedure. There are three main aspects which have had direct impact on planning through Articles 1, 6 and 8, respectively: the right to the peaceful enjoyment of possessions, the right to a fair and impartial trial, and the right to enjoy family and home life.

The formal adoption and application of the HRA (and the ECHR stipulations) has stimulated debate over how those provisions would impact on the planning system, given the way that public interest and defined planning procedures, underpinned by relevant legislation, regularly assumes priority over private (property) rights or indeed rights expressed in Articles 1, 6 and 8. Article 6 questions how the executive arm and legislators (i.e. politicians acting as planning decision-makers) can also act in a juridicial capacity over planning inquiries. This area of planning law is still subject to developing case law and precedent (see Maurici, 2002, 2003) and the HCA/ECHR still provides a basis for legal challenge if planning processes appear to contravene the Articles cited. Since 2001, there have been a number of high profile test cases concerning rights to remain on land without planning permission, specifically where the rights under Article 8 (the 'right to family life') appear to override planning law and policy. That is, where social and moral considerations have been deemed to outweigh both other material planning concerns and the standard application of existing local planning policy (see Allen, 2005; Maurici, 2002, 2003; Moore, 2010). The process of interpretation is ongoing and reflects the contingent nature of rights and their development or integration in common law. This situation differs in some respects to countries with formal and written constitutions, but they too are subject to changing legal and procedural rights which act to restructure both systems of planning and rights outcomes that are consequential to planning.

Conclusion

The examples here provide a snapshot of how rights shape planning, that is, showing how planning is a contentious and important activity in terms of allocating and shaping rights as 'benefit streams' (e.g. development rights) and how planning activity is also shaped and regulated by other civil rights. The existence of such rights and claims for rights to be enforced or interpreted in particular ways highlight the endemic conflicts and challenges that planners deal with. These are derived from arguments relating to the relative freedom that individuals have from state intervention or control and vice versa.

The issue of rights in planning links into important debates, not only about rights to participate but how such rights are exercised, enabled and encouraged by planning authorities and governments (see Chapters 9 and 10). Legal challenges to planning decisions have been a long-lived feature of planning systems worldwide, particularly given the large sums of money at stake and considering the possible environmental and social costs that some planning actions imply. This is also where different anticipations of human rights, property rights and wider civil rights combine or clash. Appreciating the role of rights *in* planning, as well as how rights are affected *by* planning, helps us understand the way that different interests behave, and why conflicts and controversies arise over planning decisions.

The concept of rights is important for planners for at least three reasons: firstly, to develop an understanding of extant rights and powers available to planners to know what is possible and legal; secondly to ensure a degree of awareness of how property rights and individual benefits are affected by planning policies and decisions; and, thirdly, to understand how planning affects, and is structured by, wider civil and human rights. The last reason is important – planners should be aware of wider changes and considerations that are outwith the confines of planning systems. It may also be argued that anyone interested in land, property, development and planning, or who is involved in mediating change, should need to understand the socially constructed and contingent nature of rights and property rights and the implications in philosophical and economic terms of altering rights distributions.

FURTHER READING

Several of the other chapters cross-referenced above will reinforce understanding of this topic, particularly Chapters 2, 3 and 16. The Freeden (1991) volume on rights is a useful general account while Allen (2005) discusses rights and the HRA in respect to property specifically. Ellis (2004) provides a good overview and a discussion of rights of third parties in planning. Parker (2002) details rights, citizenship and land in a rural context, while authors such as Pennington (2000) provide a rather heterodoxical view about how a rights-based planning could largely replace the UK 'plan-led' system. Accounts of property rights and the philosophic underpinnings of planning are also found in Bromley (1991); Becker (1977) and Allison (1975). A seminal, if general, base that many authors draw on is the Honoré (1961) text and also the Demsetz treatise on property rights (1967). Beyond this there are numerous articles and books on citizenship and rights, including critical accounts of the role of rights in society, for example Marshall and Bottomore (1992), Cranston (1973) and Blomley (1994). The planning law press is also a good source as to the legal position regarding formal rights and their interpretation: the *Journal of Planning Law* and the *Journal of Planning and Environmental Law* being main sources in this respect in the UK, along with standard English planning law texts, for example Duxbury (2009) and Moore (2010). Similar volumes relating to different legal jurisdictions exist on a country-by-country basis.

13 PLACE AND SENSE OF PLACE

Related terms: community; conservation; networks; genius loci; identity; placelessness; non-place; environmental justice; character; meaning; attachment; locality

Introduction

Arguably much planning thought and policy centres in some way around a concern for space, place and spatial relations. The RTPI's strapline of 'mediation of space, making of place' aims to convey how concerns over space and place are central to planning practice. Indeed planning activity does certainly affect the way that space and place change or are maintained. An understanding of how and why planning acts to mediate and help shape place is important because it is both a primary aim and outcome of planning activity.

While place may appear a quite straightforward idea, various elements and difficulties associated with the concept of place and in the related idea 'sense of place' are explored. This is because place resists simple definition or standardisation, and is increasingly affected by a whole range of local, global, cultural and other influences. Place-making is also a stated aim of much planning activity internationally, and given that concern for place and with community is one of several primary concerns in planning, 'place' provides a rhetorical basis or substantive justification for action. Regulatory practices extended through planning can shape perception of place, and the meaning and values that humans attribute to places. It is this interaction between the environment, in the broadest sense, and people that acts to construct 'place'. McDowell summarises this as: 'a reciprocal relationship between the constitution of places and people' (1997: 1). This can be extended further, because the process is both a mutable and mutually constitutive one in that some control or management of both people and environment acts to create a shared, particular and distinctive identity as well as wider, diffuse or organic change in cultural or environmental terms.

We argue that places are co-constructed by a wide range of factors and people, and the influences include policies, the environment and economic conditions. Planning can claim to be only one of many factors that make or shape places. Planners do need to reflect on the impact of planning and the other influences on place. The planning profession has been criticised for its past role in shaping approaches towards, and outcomes in, towns and cities which largely depended on: 'a faith in scientific, technical logic to the exclusion of ideas of place values' (Crang, 1998: 107) or, more charitably, such techniques became dominant (see Hall, 2002). The RTPI's own assertion about spatial planning highlighted above is partly explained through a statement about the main purposes and methods of planning for places:

> Planning involves twin activities – the management of the competing uses for space, and the making of places that are valued and have identity. These activities focus on the location and quality of social, economic and environmental change' (2010, n.p.).

In this view, places are seen to comprise land uses and physical forms and to have cultural and social significance, and that these together produce meaning. The statement also suggests how planning is involved in shaping and directing change in and for those places and trying to balance priorities and impacts that affect place.

We now set out understandings of the key terms in use and then focus on exploring the significance of the idea of 'sense of place' and *genius loci* (Jiven and Larkham, 2003; Norberg-Schulz, 1980). These notions have particularly influenced planning and urban design considerations and impacted on wider urban policies, particularly concerned with community and amenity in regulatory and policy planning. This account can therefore only briefly mention a wide-ranging and long-debated literature on space/place more generally. To do otherwise would be an enormously lengthy task, and one that has been attempted elsewhere. Instead we will focus in on sense of place and the way that 'place' is invoked, conceptualised and mediated by planning and related disciplines.

Space and Place in Planning

Space has traditionally been thought to be a universal, abstract phenomenon and subject to scientific laws, often viewed as a definable area. In this traditional, or Euclidean account, space is measurable, observable and limited or bounded. Efforts to define space in this way, however, have been criticised as a simplification. They are often supported by an instrumental rationality that can fail to understand the social construction of space, or value the richness of meanings and cultures of 'places' or the shrinking, folding or pleating of space that competing conceptualisations provide, as with the notion of time-space compression (see Chapter 4).

Relph (1981) for example identified four sorts or 'layers' of space, starting with 'pragmatic spaces' relating to human location and orientation within a territory and how we navigate through space. Secondly, he labels 'perceptual space' reflecting that what we experience or observe is largely centred on the self and will be an individualised experience. Thirdly 'existential space' that is informed by the culture and meanings that we perceive from place and, lastly the idea of 'cognitive space' as a combination of spatial relations or uses and interrelations between elements. Crang (1998) indicates that many planners and geographers overemphasise this latter conceptualisation of space to the detriment of the other forms. This can lead to overly simplistic accounts of place and the constituents that are important to people and community. Thus, simultaneously space is identifiable as 'place' or a series of places as spatial experiences which inhabitants, visitors or the viewers of that space value for different reasons. People respond to different features and might well hold shared aspects of place identity. This shared appreciation is often referred to as 'sense of place' (discussed later), which also links to considerations of community (Chapter 14).

The most basic definitions of place see it as a 'portion of space'. Place is often distinguished in terms of scale and in terms of the differentiation of space, such that when a location is identified, or given a name, it is separated from the undefined space that surrounds it. Tuan (1977) suggests that a place only comes into existence when humans give meaning to a part of the larger, undifferentiated geographical space and with this emotional ties and affective feelings are combined through experiences and interaction in and with that place. As such, places are seen as socially constructed and develop over time as a result of habitation, (re)development, incident and memory. Place becomes an important container and stimulus for community and affective local relations. There are different components that have been identified by geographers and planners as important potential constituents of place. These include the natural and built environment, activities and the interrelations between people that may be refracted through memories, images and indeed via policies. Tuan (1974) coined the term 'topophilia' to express the relations, perceptions, attitudes, values, and worldviews that affectively bond people and place together. This is an aggregate of a fluid and complex set of conditions and processes linked to both physical and psychological responses to experiencing place. Tuan's work also raises the possibility of individualised place imaginations and senses of place and therefore what is considered important or cherished by one is not always likely to be so for others. Changing, diverse and more transient populations may also undermine the idea of a shared sense of place. Anderson (1991) argues that such places require 'imagined communities' that opt into particular beliefs and symbols selectively and then assume a 'belongingness'. So planners are faced with a real challenge to understand and protect 'places' to try to ensure some continuities and the potentials for a shared sense of place. One resulting question for planners becomes: How to manage and understand the merits of change and continuities of place?

Changing dominant or competing conceptualisations have been explored with respect to place identity and how places are shaped by groups, activities and memory. Historical representations or heritage can become a 'metaliction'; a grand fiction deployed as a plaything or political resource for the present and be used to defend or reshape places and perceptions of place (Ashworth 1994, 1996). Places in this reading reflect assemblages of influence, including the media and competing styles or tastes and the multiplicity of groups and understandings and imaginations wrought through processes of change and interpretation.

Punter and Carmona (1997), for example, distil or simplify these considerations into three elements – 'activity, environment and meaning' – that act and interact to shape place and form sense of place. However, these components mask a number of issues including change, diversity and the politics of place identity, which produce fluid and contested places and place relations, defy reification or which are shared by all inhabitants of place. A range of disciplines have contributed to widen understandings of place. Insights from environmental psychology have informed the practice of planning and of 'place-making'. In this connection, Untaru summarises the role that places can play in people's lives:

> people think of places as the major contexts for actions and interactions within which a number of situations, scripts, and behaviour settings may occur, and these are invested with meaning, values, feelings, sensations, emotions, memories and desires. Places are therefore social constructs, given identity and value by history, and through the everyday experience of living, visiting and working in them. (2002: 173)

However, such nuances still contrast with essentialist conceptions of place that remain dominant in planning (see Hillier, 2007 and Chapter 4). A relational view of space and place emphasises connections and relations between, among and within places and spaces, and contests the idea that policies for bounded areas that focus on populations, activities and resources located within that space can be truly effective, meaningful or comprehensive. The idea that space is simply a fixed location bounded geographically is also clearly weakened by the effects of globalisation, mobilities and technological advances in terms of communications and the advent of cyberspace (see Chapter 11). The term 'hyperplace' (Harvey, 1989; Soja, 1996, 2003) has also been used more recently to reflect the impacts of global flows and of postmodern influences, which rupture stable meanings and identities and challenge ideas of the traditional or organic.

Exogenous and localised relations and diversities have been underplayed by traditional conceptions of bounded places and labels that act to reinforce this, such as 'towns', 'cities' or 'districts', and which tend to be portrayed as bounded and separate in many plans and strategies. A 'shrinkage' of the effects of space and distance on human and human-nature relations has been noted, as well as a growth of different forms of space or spatialities, notably 'cyberspace' (Dodge and Kitchin, 2001). The first has been expressed through the term 'annihilation of space', which has been noted since the nineteenth century and which accompanies the associated phenomenon of time-space compression that acts to forge new 'glocalised' relations (Robertson, 1995; Svensson, 2001). There are ongoing debates about the effects and importance of time-space compression on places (see Bridge, 1997; Massey, 2005), which warn against any assumption that space, distance and local relations have been totally overthrown by the global or network society; rather, these present one of a series of ongoing factors in the development and shaping of place. Importantly, such dynamics may affect place perception and put pressure for change and for new development in established 'places'.

Instead of focusing on debates over the relations and interactions, flows and connections across space and time that shape towns and cities, we remain within traditional parameters of thinking about place. We concentrate on the way that place is used rhetorically and in policy terms and continue a social constructivist critique. This is set alongside an appreciation of the constituents and elements that shape place and constitute the *genius loci* and the closely associated and influential idea of 'sense of place'.

Sense of Place and the *Genius Loci*

Planning has tended to consider place and place-making in terms of how to manage change in given areas and how to make new quarters or towns that can boast a 'sense of place'. Such thinking has exhibited a rationality that imposes order, rules and abstract theory onto space. Competing conceptions of place, and attempts to impose or reify place identity, have grown from a tension between the search for generalisable or even universal knowledge or claims over place. There are efforts to territorialise space on the one hand that tend to homogenise and regulate, and on the other hand a respect for local uniqueness, context and diversity which may accept partial knowledges, local autonomy and the effects of rapid socio-cultural change. These tensions reveal divergent preferences, experience and contingency in terms of changing place characteristics and the accumulation and selective use of histories to inform decisions about change. In short there is an ongoing contestation over similarity, continuity and dominant views of place versus difference, change and multiple realities that are anchored spatially. In policy terms, there tends to be a conflation of the overall experience of place and its character, which is present in the former approach.

This immediately leads to a critique of the way that planners, designers and some local or national groups have attempted to impose or 'freeze' place identity, character and sense of place through policy, marketing and (top-down) masterplanning techniques. Such efforts can result in a dominant representation of place that selectively draws on place characteristics to produce place images. Furthermore, the approach to creating new developments or urban districts has followed a rather 'cookie-cutter' approach. This has, at best, been informed by design theory, largely generated in the 1960s and underpinned by scientistic and positivist methods or assumptions. At worst, developments have been created with little regard to local preferences, landscapes and pre-existing elements of place.

Multiple or individualised senses of place may be either experienced or shared by groups or categories of people differently (e.g. residents, tourists, youth, incomers). Policies generated by planners may reflect the priorities, attitudes or perceived benefits of shaping place to suit one or more of such groups. There is an implication that sense of place may shift over time. Perceptions of place may shift over the course of time, across different times of day, or seasonally or through a year in a cyclical way as part of the habitation of place. Alternatively, the 'sense(s) of place' may need to be experienced or absorbed over a long period. Such changes and 'rhythms' of place act to constitute part of an overall sense of place for inhabitants and visitors.

Various ideas, formulas and 'toolkits' have been applied in planning, in simple terms, to emphasise *what* makes space a place and *how* to create sense of place. This has tended to be rather mechanistic and involved objectivisation and a 'desiccation' of place that has sought to break down place into components. The main terms or concepts in use considered here are *genius loci* and sense of place. Jiven and Larkham (2003) argue that they are used rather carelessly and also that they overlap. They claim that sense of place is a wider concept that can act as a container for a range of objective and subjective elements. *Genius loci* has tended towards a root in narrower, more objective elements of place. Our appreciation is that there are flaws in any attempt to impose or effectively map places from a top-down and static perspective. However, the way that these terms are used and defined are important in understanding how planning and urban and landscape designers have sought to preserve and create 'place'.

Genius loci

Genius loci reflects the notion that places may have some special characteristic or combination of factors that produce that specialness. Crang (1998) refers to *genius loci* as being an 'essential quality' which can generate an attachment to place. It is literally translated to mean the 'spirit of place', or what is unique about a place. This may be focused at the level of a few streets, a neighbourhood, town or a wider landscape. Unlike the wider conception of sense of place, below, these are more likely to be observable physical factors or constituents. The basis for the use of *genius loci* in the literature stems predominantly from work by Norberg-Schulz, where *genius loci* is said to reflect the aggregate of the constituents that relate to townscape and physical environment and how that is perceived. Norberg-Shultz (1980) is at pains, however, to acknowledge how meanings and the symbolic understanding of place also contribute to the *genius loci.* Jiven and Larkham also appear to understand *genius loci* as the 'the sense people have of a place, understood as the sum of all physical and as well as symbolic values in nature and the human environment' (2003: 70).

Familiar features of a landscape or a neighbourhood are often fiercely defended as they become linked with cultural or personal identity. More prosaically they may be identified as positive attributes that help maintain the economic value of private property, such as nearby green space or an architectural style. Thus, as the townscape changes the *genius loci* subtly alters, and as the population changes, visitor numbers grow, or perhaps as place marketing is initiated, a similar process of change in terms of the perception of place is likely to occur. Indeed such change may be resisted by some or otherwise 'managed' by planners. Such change may not suit certain interests and provoke conflict, with planning applications and the drafting of planning policies providing a key locus for such tensions to be played out.

Sense of place

There is often a conflation of the terms sense of place and *genius loci.* Agnew (1987) points to the 'local structure of feeling' as being an important part of sense of place and this is common also to *genius loci.* We prefer to use sense of place as it encapsulates physical or observable factors as well as intangible and personal perceptions and values of place. It also becomes confusing to deploy both terms when they clearly have considerable overlap. For both concepts, particular buildings, sites, artefacts, aspects of the natural or created environment, historical figures, or recorded actions and events that have been committed to memory or emphasised, may create a shared understanding of place. A local degree of consensus or assertion of the relative significance of elements of place character and identity may have built up over time, possibly through the actions of local and national elites. Areas that are said to have a strong sense of place are likely to have an identity and character that is experienced and enjoyed by local inhabitants and visitors. The sense of place may be enhanced, reconstructed or otherwise portrayed by novelists, musicians or artists, too. Some places may have been accorded some special status to reflect some of these features and this may act to reinforce that particular construction of place and sense of place (such as CAs or protected areas, e.g. NPs).

Sense of place implies some potential for emotional attachment or reaction and designers have tried to replicate or retain features of the built environment that are likely to provoke positive feelings, or maintain certain continuities with existing place identity. Emotional ties may include relationships with particular features, stories and folk-memory. These can relate to personal memory of places, often associated with events, to people and interaction with the place as a setting for experiences (for example, where a lasting personal experience takes place such as the setting of a childhood holiday, or the street corner where one experiences a first kiss). Understandings of sense of place recognise that place-making or shaping can be instrumental in community development and shaping some aspects of quality of life. Much of this thinking has influenced parts of the UK government's *urban renaissance* agenda formally launched in 2000. For example:

> The urban environment can be harsh and intimidating or it can encourage people to feel at ease. It can be impersonal and make contact between people difficult or it can foster a sense of community. (ODPM, 2000a: para. 4.2)

And asserting that:

> In England we have long had a tradition of creating towns and cities of quality and beauty, places that can bind communities together. (ODPM, 2000a: para. 4.4)

There may be contested or multiple overlaid appreciations of place with different elements or signifiers being recognised or prompting different reactions. Rapid or wholesale change can 'rupture' continuities of place and where change, or specifically new development, is proposed or permitted, urban design solutions are often sought to mitigate the effects of change on place. The approach against mixing of land uses in some countries has not helped to achieve a dynamic, meaningful and pluralist urban environment. One outcome is the lack of a mixed or large enough population in city centres to engage effectively and contest development proposals and a concomitant growth of so-called 'clone towns' (Simms et al., 2005), lacking diversity or, some might say, a sense of place.

This presents a particular challenge for planners in assessing the effects of change on particular valued associations and elements of place that contribute to locally and more widely held appreciations of place. Ideas of some shared or uniform sense of place may smack of a top-down imposition and reification of place but often they are seen as the only practicable option in producing new spaces/places that will be 'adopted' by populations. The process of achieving such outcomes has been the source of much debate in planning. A dominant feature has been discussions around a perceived need to shift towards a communicative rationality and forms of collaborative and community based or neighbourhood planning in order to co-design and broker new development.

As such, the construction of the concept of place has often been shaped by the aims of its promoters, with planners trying to build up a shared experience of place through design interventions, aiming to distil shared appreciations and also to 'protect' the physical environment. A more fluid and diverse society with attendant attitudes poses a series of challenges for policymakers to achieve outcomes that foster a widely held sense of place. It is likely that a lack of understanding of place dynamics, an underdeveloped understanding of diversity, community perceptions and a uni-dimensional conceptual awareness has acted in various combinations to produce misdirected policy and resulted in many developments that appear formulaic or insensitive to wider characteristics or valued aspects of place.

Sense of place tends to convey a positive connotation, yet there will be negative place image and a lack of sense of place that plays counterpoint to this. A 'negative' sense of place may include feelings of sterility or exclusion, or be typified by architecture that can be oppressive or characterless. This idea of negative place image is partly expressed in the idea of 'placelessness' (Relph, 1976) and also the associated idea of 'non-places' (Arefi, 1999; Auge, 1995). The latter was coined in reference to modern spaces such as shopping malls and airports. In these cases the authors bemoan the uniformity, lack of shared connection to history and use, and in some cases just a lack of meaningful character that people can connect with. This has been part of a long-running argument and justification about the use of particular materials and forms of architecture for housing and other land uses, and in attempts to shape places so that they are more conducive to a positive sense of place. Indeed physical improvements are often the first step taken by public authorities when attempting to regenerate areas, although in more recent times a recognition of more integrated action in social and economic terms is apparent in policy, certainly in the UK. Wholesale change can intimate a perceived normative 'failure' of place-making that requires radical intervention (for example Cochrane, 2007). We now turn to examine this a little further by discussing how place is a key motive for urban and landscape designers.

Urban Design and Planning for Place in Practice

Urban designers tend to think in terms of the manipulation of the physical environment, land uses and the relationship this has with the individual. As mentioned, many urban designers identify a trio of elements that shape place and place perception: environment, activity and meaning. In the urban design literature, sense of place is regarded as a condition that makes an environment psychologically comfortable. It is seen as an aim in designing and arranging the physical environment to produce positive social and cultural behavioural outcomes. There are several key aspects to this, including the legibility or understanding of place which is linked to the perception of the visual environment. Another dimension connects place and person with activity and the compatibility of the setting with human purposes, meaning essentially that the design should be appropriate for the activities taking place there.

Kevin Lynch's (1960) influential work evaluates place by assessing how territories are marked and whether the 'transitions' from one space to another are clear – that is, how places are demarcated and in some way differentiated. This work implicitly involves imposing and 'designing' place and setting place boundaries in a rather top-down way. Urban designers may be interested in whether the desired range of behaviours and groups are provided for and how well users understand and agree on the meanings and boundaries of those territories and attempt a more collaborative approach as with, for example, 'Planning for Real' processes and design charrettes (see Condon, 2007). More importantly, it is argued in urban design theory that the identification of places and their organisation not only allows people to function effectively but can also be a source of emotional security, pleasure and understanding. These latter ingredients relate to memories, emotions, sensory perception and feelings that are generated or triggered through the environment. Environmental psychologists often refer to these features as 'environmental cues' that act to prompt actions and reactions (Bonnes and Secchiaroli, 1995).

Cullen in explaining the idea of 'townscape' highlights how physical and observable components have been given priority in modern planning: 'the elements that go to create the environment: buildings, trees, water, traffic, advertisements and so on, and to weave them together in such a way that drama is released. For a city is a dramatic event in the environment' (1961: 7). Designers look to other deemed successful places, both old and new to replicate features and juxtapose physical elements to reproduce that success; to make and copy 'good places'. In many cases such interventions are often made to mitigate insensitive or poorly thought through new developments and break with features of adjacent places. The urban design literature is dominated therefore by technical and other guidance for planners to replicate and 'bolt together' apparently meaningful or conducive elements to create place. Yet this can belie a danger in the creation of hyperreal or 'inauthentic' places. Examples of such hyperplaces are said to include 'historicist' or 'New Urbanist' developments such as the settlement of Celebration, Florida, in the USA or Poundbury, Dorset, in England (see, for example, Molotch et al., 2000; Bond and Fawcett-Thompson, 2007).

Alternatively the use of technical and top-down approaches can be seen in a more sinister light as attempts to segregate or divide populations (see, for example, Yiftachel (1998) and Flyvbjerg (1996) on the so-called 'dark side' of planning). This process is of course hotly contested, particularly in the light of ethnic and religious tension and calls for new spatialities to serve multicultural and integrated towns and cities.

Place and urban design 'principles'

As mentioned, attempts to design-in and create 'place' have been distilled into various codes and guides for practitioners. As part of this approach, many urban designers have translated notions of 'good place' into design principles in order to guide local planning policy and decision-making and to shape the physical form of new development. An early example is that of the Essex design guide produced in 1973 (see Essex County Council, 1973; Goodey, 1998). The work from different epochs shows that a relatively consistent set of ideas has been imposed (or has emerged, depending on the openness of the process to different interests) to structure the planning of place dimensions. There have been several variations of 'Good Place' design principles promoted in different eras of planning guidance in England though. CABE (2001) details a shortlist of seven such principles, listed as: 'Character, Continuity and enclosure, Quality of public realm, Ease of movement, Legibility, Adaptability and Diversity', while Carmona et al. (2003) discuss the development of thinking around place-making principles over time, showing different elements and emphases (see also Bentley et al., 1985; CABE, 2000; Madanipour, 1996; Morris, 1997; Rowley, 1994). These guides and principles serve to provide a steer or formula for planners, developers and others in seeking to produce developments with a better chance of success in terms of place and community. A good source of local design guides is available on the Resource for Urban Design Information (RUDI) website (http://www.rudi.net/tags/udl_content).

Conclusion

This chapter has indicated that sense of place may not be shared or static and the *genius loci* may not be universally recognised or agreed. The use of different terms such as *genius loci* and place character or identity, alongside and interchangeably, with sense of place does perhaps unnecessarily complicate things. Jiven and Larkham (2003) attempted to provide some clarification about differences in the terms and how they should be deployed with only limited success. *Genius loci* appears to refer to more observable factors of place such as landscape, architecture and land uses, while sense of place overlaps with this and also includes the personal and intangible factors that inhabitants or visitors may share. The assumption that such an identity or character, or set of definable attributes truly exists and can be protected, created or reproduced has predominated in planning practice. However, it is argued that place and sense of place is formed through a highly complex set of interrelationships, only some of which are understood or controlled by planning. It may be that dominant authority and groups within and outwith a locality will attempt to maintain particular assets and characteristics and rhetorically construct place and identity for their own ends. This can, at worst, be elitist and exclusionary, but may also provide benefits in cultural, economic and environmental terms. As such, there are many examples where a selective and partisan use of sense of place as a rhetorical tool has been used to prevent change. In among such constructions, planners attempt to manage physical change and calls to integrate policy as part of a shift towards 'spatial' planning. This implies that a new sensibility, approach and methods will be needed towards a place-shaping agenda that aims to improve quality of life and 'liveability', as well as promoting environmental justice and accessibility to services. The speed and type of change, and how change is mediated, are also seen as important in terms of eroding or maintaining some shared sense of place and in retaining some continuity with the past.

Despite the above, the predominant deployment of sense of place in efforts to manage place has become largely a preserve of the urban design process and the approach, pioneered in the 1960s, towards somehow essentialising what 'makes a good place'. Although more subtle and nuanced approaches have been developed more recently (e.g. Carmona et al., 2003), overall these approaches have entailed breaking places down into bits, features and elements. This is intended so that planners might reproduce places through the use of architectural technique and the application and juxtapositioning of deemed important elements. The collaborative critique discussed elsewhere, along with the cultural geography contributions mentioned here illustrate how such identikit approaches may only provide part of the formula needed to produce and maintain liveable cities, or to encourage community – particularly in a diverse and changing society. The environmental psychology perspective, provided by early authors such as Norberg-Schulz (1980), indicates how the person-centred and individualised perception of place is important, and this angle also highlights the role that the physical environment has on people (e.g. Bonnes and Bonaiuto, 2002). Both have obvious repercussions for methods and processes of formulating policy and design decisions, including more communicative and collaborative processes.

FURTHER READING

There is a plethora of sources and commentaries on place and sense of place which can be divided into several parts, including the wide theorisation from human geography, sources from environmental psychology (see Bonnes and Secchiaroli, 1995) and the more technical or practical literature derived from urban and landscape design. Cresswell (2004) provides a good concise account of the idea of place in human geography as does Crang (1998), and Castree (2003) also adds to this. Massey (2005) includes some thoughtful material on space and place, while the volume edited by Hubbard et al. (2004) provides a wide set of essays (totalling 52 chapters) that set out the work of many of the authors who have engaged in thinking through the meaning and import of space and place. Similarly, the reader on place edited by McDowell (1997) is a useful text to explore different angles and appreciations of place. The broad critique of competing concepts of space and place as applied to planning is presented nicely by Graham and Healey (1999), while also linking these to new appreciations of space-time and the relevance to planning practice of relational place and space discussed in Chapter 4. The Jiven and Larkham paper (2003) is useful reading as it challenges the way that urban designers have used place and sense of place and was itself prompted by Ouf (2001). However, to understand the way that place has been broken down into constituent elements by urban designers, Lynch (1960) and Cullen (1961) are the basis for much design theory and are still used by design instructors. For more up-to-date urban design handbooks see Carmona et al. (2003) and Punter and Carmona (1997); and a good review of urban design is included in Madanipour (1996). The sustainability agenda and creation of lower impact development has also been considered by Carmona (2009), who poses 10 'questions' for developers. CABE's *By Design* (2000) provides an influential handbook arguing for 'good design' as part of the UK's 'urban renaissance' and was created to sit alongside housing policy guidance in England with a focus on 'place-making'. In policy terms, the English policy guidance PPG 15 on the historic environment has been replaced by PPS 5, but this is useful as a policy platform for students (and planners), and both are worth reading for the assumptions they contain on sense of place. There are numerous sources to investigate community planning techniques, including http://www.communityplanning.net, which also explains the 'Planning For Real' approach pioneered by the Neighbourhood Initiatives Foundation. The various design principles expounded over time are traced in Carmona et al. (2003) and in the more recent expressions of design guidance listed on the RUDI website (http://www.rudi.net/tags/udl_content).

14 COMMUNITY

Related terms: cohesion; quality of life; participation; identity; sense of place; amenity; networks; capital

Introduction

Community is a well-worn term that has been used and misused in public discourse and broadly across the social and political sciences. In planning terms much of the activity of planners is justified as being in the public interest (see Chapter 9) but more and more the notion of community is attached to a variety of planning processes, policies and actions. This common association of planning activity to and for community as both an end and a stakeholder group justifies an exploration of the term and its relevance for planners and in planning practice here.

Community was seen as a political ideal in the ancient world, where citizens would participate in public affairs as part of the community. The concept has developed such that 'community as belonging' has come to be viewed both as a past state and as a desirable aspiration. Hobsbawm pointedly observes 'never was the term "community" used more indiscriminately and emptily than in the decades when communities in the sociological sense became hard to find in real life' (1994: 428). Even more pessimistically Bauman (2001) indicates that predilections towards recovering or developing community ignore the likelihood that it never existed in the first place. Taking a rather different tack, Delanty (2003) highlights that early conceptions of community were posited in *opposition* to ideas of the state or to wider society and reflected instead the desirability of a strong 'inner world' that rejected wider society as being in some way undesirable or degenerate.

Such critical views are balanced by thinkers who have pointed to examples of community in practice and to desirable constitutive elements of community. The philosopher Immanuel Kant identified, over 200 years ago, a key element in community that appears to endure even in postmodern times: the characteristic of *communication* in communities and the importance of relations in developing and maintaining different communities (see also Chapter 4). This overlays any preoccupation with locality or community as a symbolic, normative or utopian ideal, and therefore still resonates with us today.

Definitions and Uses of Community

Traditional use of the term community has also regarded shared relations and groups of people within a locality, as its central defining characteristic. This usage of the term can obscure as much it enlightens though, serving a range of political purposes. Conceptualisations of community vary widely, but the term is often left unexplained or ill-defined when used. The scale of community can equally be unclear or left open, with some utilising the term to refer to specific groups, others to represent whole populations, or otherwise as a shorthand for an intervention that will benefit 'the community' at an undefined scale or type. Community has also been used to convey the idea of an idealised state of harmony, rather than reflecting an actual state. In these usages community is generally misused to depict an 'arbitrary inclusivity' and to assume homogeneity and some sense of shared values and attitudes that may be misleading. In this way community may be propagated to instil aspirations for idealised communities, or to provide a positive spin to the actions or claims of a range of stakeholders who may play direct or indirect roles in spatial planning.

Many have lamented the breakdown of community, citing this as a reason for a number of social ills. This assumption has underpinned a range of actions and behaviours on the part of governments and planners over recent decades. Healey highlights how 'politicians, citizens and planners often talk nostalgically of community as a time when everyone living in an area knew and trusted each other' (2005: 123). This places the roots of the normative idea of community as one where local relations were based on shared activities, particular common understandings and congruent values. This reflects the post-hoc idealisation of community distained by Bauman. Williams (1983) provides a good overview of the different definitions of community in relation to interaction and shared purpose, as well as noting how community is sometimes expressed in opposition to 'outside' influence or threats of change. In this context 'community' may only be apparent if a challenge is posed. In planning terms this is seen quite clearly when protests and objections are made and 'community' action groups seek to lobby and challenge developers and planners. In this situation there is likely to be an overlap between place and interest, although the representativeness of those speaking on behalf of the community will be uncertain.

There are several characteristics shared across competing definitions of community. Local and common concerns, as well as functional relations, can act as unifying factors. Individuals may choose to be members, or be assumed to assent to membership (see Anderson, 1991), others assume community as comprising inhabitants rather than 'members' that actually share characteristics or attitudes.

Identity is a key element of community where the individual shares some common concerns or experiences and this also brings into view people–environment relations and concerns with place discussed in Chapter 15. The presence of shared identifiers, however, does not guarantee mutual identification and membership. It should be recognised that individuals may have numerous facets of identity and self-identify with multiple communities that reflect their different ties, interests and attitudes. Residents of a particular geographical locality may be viewed as part of an *area-based* community and are sometimes referred to as 'propinquitous' communities. These may be conceived at various scales from streets to neighbourhoods, villages, towns or districts. We can discern a clear division here then which rests between area-based understandings of community and those based on *interests* and shared concerns or values, and therefore do not necessarily equate to particular bounded spaces. A loose view of community as concerning common characteristics, or people sharing an 'attachment' to place, has become more accepted, and bridges across those two categories. It should be conceded, however, that any idealised view of community is unlikely to be an accurate representation.

Indeed assumptions about positive attributes of community ignore the numerous commentators who have pointed to the negative features of community membership (e.g. Bauman, 2001; Crow and Allan, 1994; Williams, 1977). This can include intra-group conflict, repression and a lack of privacy, which may produce a stifling and normalising effect on individuals. Crow and Allan (1994) point out that community membership may not be chosen and the experience of 'being in a community' may give rise to uncomfortable tensions. Bourdieu's ideas of *field* and *habitus* assist here, indicating how relations act to structure behaviours and how understandings reached and conflicts mediated by communities can be understood (see Chapter 13; Bourdieu, 1994, 2002; Jenkins, 1992; Hillier and Rooksby, 2002) with the norms and sanctions exercised within communities acting to regulate behaviours (Coleman, 1988).

Stable and conducive relations are less likely to exist in a postmodern society where people are more mobile, diverse and less constrained in terms of expressing difference. Individuals have numerous other opportunities, distractions and resources that allow relative independence from place and local relations or ties, and people are more able to ignore local or group sanctions (Fischer, 1982; Ghezzi and Mingione, 2007). Webber (1963) pointed out that this is not a new process and that socio-economic-technical changes jeopardise traditional views of community. This fluidity in group identification highlights the difficulties in relying on community development as a means to achieve more liveable places. Social change has undermined strong or dense local relations and also legitimates actions to plan for diversity and attempts to create a more inclusive planning to explore the spatial and physical needs of such populations (see Greed, 1999; Hastings and Thomas, 2005; ODPM, 2005a).

Unanticipated strains and pressures on land uses and other flows can actually serve to refigure and in some cases erode established area-based communities as people become disembedded or refigure their relations in heterogeneous ways (Murdoch, 1998). Communities of interest and their ability to interact are part of a wider shift in behaviour patterns and flows, for example with online communities, leisure-based groupings as well as business communities interacting globally, and given that people tend to be far less tied to particular locations in order to socialise or work. Attention given towards interest-based community (see Crow and Allen 1994) has shown how there is significant shift of impact on the way that space is used, how people travel and use resources, as well as transforming former land uses and loci of social interaction, such as churches, schools and festivals. Indeed the idea of communities of interest and their role in society has come to the fore, presenting a number of challenges to traditional policy assumptions.

It is also contended that communities of interest not only operate across spatial scales, but also engage with local governance networks through representative or special interest groups and have spatial needs or claims. In a more diverse population, there will be multiple layers and multifaceted communities that require different methods of engagement. Acceptable solutions may be more difficult to find. The notion of the 'political community' therefore acts as a bridge between fragmented populations and the erosion of place or locality as the main determinant of community. In some cases 'communities' may cohere around proposed developments or policies drafted for particular areas or sites, or instead constitute part of semi-formal governance arrangements that are maintained locally (see Doak and Karadimitriou, 2007; Raco et al., 2006). This may require or be a product of innovative arrangements for community involvement where resources are tight or where planning proposals are anticipated to be controversial among some groups or segments of a population.

In practice planners simultaneously interact with and shape 'community' and 'communities' in their own jurisdictions. There are two planning dimensions here in terms of what is left behind or becomes less important and in the challenges of new dynamics that require different solutions or regulatory positions and in which planning policy can play a significant role (as discussed in Chapter 11).

The analysis of community as a reification, convenience and simplification has not necessarily diminished enthusiasm for community as a central consideration and a way of labelling of stakeholders in spatial planning. Tensions can therefore exist between top-down and bottom-up notions of community and appreciations of community in calculations of the 'public interest'. The idea of community provides rhetorical support for numerous policies and actions; and indeed government have found that the term community can be used liberally to provide an attractive label for a great variety of public schemes and initiatives. This brings us more directly to consider how notions of community and planning come together.

Community, Politics and Planning

Despite a shaky basis in past conditions and present idealisations, the maintenance of community is seen as an important legitimating idea in planning. It has been a substantive goal for planners to seek to assist in developing and maintaining (local) communities as a means of social cohesion and quality of life considerations. One component of community is claimed to be fostered through the appropriate organisation of land uses and use of design principles, as well as a number of different activities and institutional arrangements beyond planning. Community(ies) is also increasingly seen as the key beneficiary and co-partner in planning and efforts to engage with (multiple) community (groups) are widely discussed and attempted in many countries. Overall this view of community reflects an imposed generalisation of community and is closely associated with justifications for planning in the 'public interest' (Chapter 9). This may be seen as providing part of a discourse of legitimation for planners and politicians in order to help justify decisions and to implement policy. However, what is community and why is the term so widespread and apparently powerful?

The label of community has been pressed into service for a number of purposes and can imply different motives and meanings. Not least of these include numerous governmental schemes, with 'community' providing a positive, convenient and aspirational label for a range of initiatives and their objectives. In planning it has a justificatory implication, largely because of the claims made for planning to serve and engage with 'communities' and to help develop a 'sense of community'.

Community has a strong positive connotation and is used in public discourse almost exclusively to represent an aspirational state or to assert a favourable social condition. Perhaps due to this dominant cultural understanding it has been co-opted by politicians and policymakers. Where government act 'on behalf of the community', the usage and definition is generally loose and could be replaced by less appealing alternatives such as 'public', 'state', 'local' or perhaps as 'targeted' policy. It often becomes a means of euphemistically communicating and labelling and grouping a wide range of groups, scales and policies (Paddison, 2001). In this usage questions are raised of motive and of attempts to control and incorporate interests (see also Cochrane, 1986, 2003, 2007).

According to Delanty (2003) community is increasingly relevant when considering the communication age and the implications that this has for understanding types and dynamics of community in an era of 'detraditionalisation' (Heelas et al., 1996) or where disembeddedness (Ghezzi and Mingione, 2007) occurs; whereby identities and relations are fluid and less likely to be based around social norms and place-identities formed or inherited as part of a dominant tradition (see also Friedmann, 1993; Sternberg, 1993). This is another reason why a perceived loss of community becomes linked with concerns over 'sense of place' (Chapter 13).

The broader, weaker usage of community that tends to be assumed can obscure difference and diversity. Cochrane succinctly illustrates how some see the possibilities of this happening where community politics and community engagement by planners and others are applied cynically or loosely:

> as if it were an aerosol can, to be sprayed on to any social programme, giving it a more progressive and sympathetic cachet. (1986: 51)

The concern being that efforts and examples where community is invoked or claimed to be engaged can often be façadist or aim at incorporation of sections of the community rather than genuine participation. This laden term invokes an imagined cohesiveness and mutuality on behalf of members or populations within set bounded areas such as streets, villages, towns, or for particular groups within localities. In this sense it more clearly refers to groups that share some common relations and identity. Community can therefore be a self-referencing tool for groups wishing to define themselves in a characteristic way. Community is also used to reflect a bottom-up process which may be posed in opposition to authority or outsiders. Conversely, it may be imposed and applied from above by politicians or researchers, or indeed planners. Where attempts to impose some abstract notion of community are made, or where some other threat to a group or area is perceived, then the term 'community' can emerge oppositionally, for example through a local action group.

Concern with community in planning terms centres then on at least three main aspects. Firstly, it is used as a useful political euphemism and a simplification for the neighbourhood scale. Secondly, as justification for action and for an imagined body to be mediated for 'in the community interest'. The third relates to groups acting within the embrace or under the label of community and that look to proactively influence policymakers as a stakeholder group (see Chapter 9). In the first case community is seen as a convenient shorthand for the local scale for planners and economic development agents to seek views and engagement and so to legitimate plans, policies and bids. Community is also circulated to serve as a regulator and represent an aggregate of relations that requires planners and planning policies for support and legitimation. In this view 'community' is used and represented in an overlapping manner: as local population and as end state or aim. Both are often needed in order to underpin both local–national decision-making processes and outcomes in a variety of ways. This is seen in the literature on urban regimes and community politics, for example (cf. Stoker, 1998; Stone, 1989), where local populations are consulted or 'represented' but with numerous claims that they are in some way exploited, or that consultation is limited by institutional design and available resources. Efforts to engage have also been dogged by a suspected widespread culture of elitist and traditional politicking that can militate against local empowerment.

Numerous attempts have been made to maintain or encourage the development of communities and there is a strong history in theory and in practice where the idea of 'the community' has been seen as a justification for action. While the above criticisms of uses of community as justificatory tag exist, there are examples where planners have sought to act more directly on behalf of communities and segments of communities. Advocacy planning is one form where authors such as Davidoff (1965) and Gans (1969) noted how planners could act to support disadvantaged groups or other minorities that tended to lose out in planning decisions and consultations (see Campbell and Fainstein, 2003). Advocacy planning is a proactive intervention that aims to help rebalance access and power relations to achieve favourable outcomes for those groups. This idea led to reassessments of public and community involvement in planning and augured shifts towards the collaborative planning model and a tradition of acting with community groups both in terms of formal planning process and wider community development efforts in a number of countries.

Communities, and the term community, have been brought into planning practice in a direct way in the UK. Governments have recognised a wider disconnection between people and state (and planning decisions, outcomes and local people) and sought to broker community cohesion, often targeting specific areas or groups as 'failing' communities, largely in urban areas. Community is also invoked in relation to engagement in the planning process more generally, with numerous efforts to find appropriate means to encourage the public as 'community', in taking up participation opportunities (Brownill and Carpenter, 2007; Brownill and Parker, 2010; Doak and Parker, 2005; Kitchen and

Whitney, 2004; RTPI, 2007). Despite rhetorical and suspected façadist efforts to understand and engage with community, there is a more general need for planners to better understand the complex relations and histories, feelings and attitudes of populations as part of a more reflexive and humanist planning. Two examples of policy vehicles and associated processes that rhetorically rely on and invoke community in planning are aired below.

Community involvement in planning

The term community involvement has been coined to express how a widened engagement with planning and civic concerns has been promoted, including the encouragement and identification of different communities or groups within neighbourhoods and districts (see Campbell and Marshall, 2000; Haus et al., 2005). This also reflects how diverse constituencies that exist in localities have not necessarily been well served by planning. This is despite many governments claiming to want to foster a more inclusive planning. Formal community involvement is usually sought at set stages in the planning system and in delimited circumstances in relation to the planning process. In England these steps were outlined in statements of community involvement (SCIs) linked to each planning authority's LDFs under the 2004 revisions to the planning system (see Brownill and Carpenter, 2007; Doak and Parker, 2005; ODPM, 2004b) and guidance was provided by central government and others (DCLG, 2008; RTPI, 2007). The, albeit patchy, evidence suggests that many local planning authorities (LPAs) do the minimum to engage with the 'community', partly due to the lack of specificity about who and how they should involve and more importantly who they *must* involve. Planning also still suffers from ongoing limitations derived from informational deficits, resource constraints and asymmetrical power relations across different stakeholders (see, for example, Moulaert and Cabaret, 2006) which exacerbate the situation. In some instances local planning authorities may actually create 'convenient' obstacles to involvement and effectively disincentivise engagement by using standard consultations and methods of involvement that provide 'tick-box' solutions to stipulations to consult or involve that are handed down from central government. For more on methods and approaches, as well as critiques of community involvement (see Action with Communities in Rural England (ACRE), 2007; Brownill and Parker, 2010; Cochrane, 2007; RTPI, 2007; Wilcox, 1994).

What is clear is how community is simultaneously seen as a core concern for public sector planners and a term used to promote a range of efforts unified in a common purpose to address social problems. Similarly, it is also seen as a means to create opportunities for involvement in decision-making and ensure that people are adequately informed, voiced and understood. The way that such efforts are operationalised, however, often leaves much to be desired.

Community strategies in England

Attempts to engage and widen the subject matter of concern for spatial planning, as well as the different likely interests and concerns emanating from the 'community', are reflected in the linkage between attempts to forge a new local governance or 'new localism' in the UK (see Cochrane, 2007; Corry and Stoker, 2003; Imrie and Raco, 2003; Parker, 2008) and spatial planning (and which is an ongoing project). This has involved a widening of aspiration and exhortations to engage and work in partnership with stakeholder groups across the local authority policy range and implicitly that community will be represented through local partnerships.

The linkage of community strategies (rebranded post-2006 as 'Sustainable Community Strategies' or SCSs) with LDFs in England provides us with an example of attempts to link diverse aspirations of different groups and across different policy silos to the new spatial planning and in the wider context of shaping sets of governance relations at the local level. Community strategies were conceived as ways of brokering action and partnership working in areas of disadvantage – originally targeted at the 88 most deprived areas of England in the late 1990s as part of the government's Neighbourhood Renewal programme. These were intended to reflect community aspirations and involve the community in putting together priorities for action. Doak and Parker (2005) illustrate how the thinking behind partnership working and local governance presupposes and requires active community engagement and review how government initially saw SCSs working. This was despite ongoing difficulties with achieving deliberative, quality and meaningful interaction with representative communities and other stakeholders in planning and local governance. Overall SCSs were supposed to reflect the needs and aspirations of 'communities' at the local level and guidance was prepared to support stakeholders in developing and championing neighbourhood (community) leadership. SCSs were also conceived as a means to organise and bring together private, public and voluntary sectors and to engage with communities more directly to shape priorities; both ostensibly and eponymously for the local community. Since their first inception explicit linkages and now a borrowing from the SCS process has been taken across to formal spatial planning but with uneven success (Brownill and Carpenter, 2007; Darlow et al., 2007; Lambert, 2006; Tewdwr-Jones et al., 2006). However, what is clear is that a gradual shift towards encouraging 'active citizenship' that may help lead to more interest and engagement with policy and planning is being developed and emphasised by the current coalition government in the UK, which wants to develop community empowerment aiming towards and 'creating communities with neighbourhoods who are in charge of their own destiny, who feel if they club together and get involved they can shape the world around them' (Cameron, 2010: n.p.).

Conclusion

Together the examples help to indicate how government and the planning process use and rely on the term community and how local populations are often reified as community for different political and information-gathering purposes (as well as a basis for political legitimacy, see Chapter 9). It is also apparent that any essentialist or 'core' idea of community has been significantly eroded, yet it is still seen as an organising label both for planning in connection to public engagement and bottom-up processes of evidence gathering, mediation and governance relations. The notion of building and maintaining an environment that enables local shared aspirations and views as an achievable goal is one which should be subject to a degree of healthy scepticism even if one may consider such ends laudable.

Community remains important therefore in signifying a desirable aim or state that planning claims to strive towards, despite the obvious difficulties both with the normative concept, the means of achieving cohesive community and the difficulties associated with developing planning systems that can cope with diverse needs and aspirations of fragmented populations. In conclusion, community is largely about meaningful relations and therefore community in planning terms is a twofold desired substantive aim and a factor of process – to strive for a deliberative and inclusive planning allied to a more consensual effort to develop and manage the physical and social fabric of society.

FURTHER READING

A good overall account of community and community in relation to local governance is found in Imrie and Raco's (2003) edited volume, which covers significant ground and illustrates how community is invoked and involved in urban policy. Bauman (2001) and Crow and Allan (1994) also provide wider discussions of the concept and application of community and Delanty (2003) similarly covers the history and theoretical foundations of ideas of community. In terms of community planning there are numerous papers and books devoted to this including Wates (1990). The UK government's approach to community engagement in planning, and local plans in particular, was set out in numerous forms but PPS 12, redrafted in 2008 (DCLG, 2008), provides a good insight into formal participation and the RTPI's guidance note No. 1 (2007) for planners on community engagement is illuminating. Another addition to the literature is Cochrane's (2007) book on urban policy that indicates how community can be and has been used as a legitimating idea to mollify and enable policies and proposals to be accepted.

15 CAPITAL

Related terms: social capital; environmental capital; cultural capital; human capital; resources; relations; networks; embeddedness; community

Introduction

Many planners and commentators on urban change in the USA and Europe had begun to appreciate the limits of land use and physical planning by the 1960s, with dominant preoccupations with design and built form, along with attempts to rationally organise land use. Some had already begun to engage with insights and theories being developed more widely in the social sciences, for example Jane Jacobs had become certain that for planning actions to more effectively generate improvements to quality of life, then the social and environmental qualities of places had to be better understood. These would need to be reflected in processes, policies and planning decisions. At this time Jacobs (1961) is credited with coining the label 'social capital' to encapsulate and highlight social interaction as a key reason for places functioning and for communities to cohere. Her observation was that settled localities relied on interpersonal relationships and trust which led to a 'feeling of community' (see Chapter 14). It was claimed that self-regulation of behaviour was important and that this should lead to safer and more 'liveable' places.

The term capital is unpacked here and extended from the label or metaphor of capital famously elevated by Marx. Various types or facets of capital have since been elucidated and discussed in relation to social structure and to human–environment relations. These are introduced to provide an overview of how capital(s) are used and affected by, or that structure, planning practice and outcomes. A large volume of work has been produced around capital in its different guises and we provide a brief account of the different forms being debated. In essence we aim to make more explicit how planning is actively involved in shaping and being shaped by various concepts and resources that draw on the term capital. This covers financial investment and economic capital, and also the range of capital labels developed to connote other 'resources' or reservoirs of socio-cultural-environmental value.

Capital and Capitals

Capitalist economies rely on economic or financial capital for their operation and many planning systems operate in reference to and act to shape and direct economic capital flows. The Marxist position that economic capital acts as a structuring force in society presents a somewhat compelling argument for the inclusion of the concept here but there are wider reasons too. For example, the way that property development requires economic capital, that is, financial investment for it to proceed. This means that an understanding of the decision-making criteria of developers and institutions relating to viability is an important part of the role of planners. Equally, there are wider issues about the appropriate use and management of resources in the process of accumulating and converting capital and which form part of the evaluation of policy options. This may mean reflecting on how, according to associated discourses about capital forms, the presence or absence of different forms of capital may affect places and spaces.

Economic capital as conceived by classical economists such as David Ricardo has it that capital is any form of wealth capable of being employed in the production of more wealth. Others consider that a good broad working definition should point to capital as a resource, for example: 'a stock of resources that gives rise to flows of goods or services' (Ekins et al., 2003: 166). In classical economics, capital was seen as only one 'factor of production' because the term excluded land, labour and entrepreneurship, which traditionally constituted the other factors of production. Capital came to be seen as assets that represented wealth and were capable of being used in production. In this view natural assets such as minerals, trees or land were therefore also understood as being elements of capital.

For Pierre Bourdieu capital was a useful metaphor for resources accumulated through differing forms of labour and 'investment'. This form of analysis drew from Marxist thinking, where the notion of capital in its narrower economic sense had dominated much of the thinking in sociology and where it was used to build a whole theory of society. For Marxists, economic capital is the source of power and control in society, and this was, predominantly, considered in terms of financial capital and the control of resources (such as land and other means of production). This has since been excavated and explored, with economists, sociologists, geographers and planners extending the field of economic development and sustainable development studies. They have sought to understand and promote economic activity of different types in different places and crucially to develop methods of more fully accounting for resources. This has led to a diversification of labels used to mark a range of capital forms and hybrids and a growing recognition of the role of spatial planning in using, shaping, developing and managing forms of capital.

Sociologists have offered understandings of how society worked and many others were engaged in developing theories about the resources and cultural values that societies have attached to particular behaviours, language and the built and natural environment. There was a recognition of the limits of traditional or neo-classical economic theory which prompted thinkers to extend conceptualisations of resources. The extension of the metaphor of capital was logical to assist theorisation about both the relationship between types of resource and why and how they were valued in society. An important element of this was a commentary about the role and importance of 'social capital' (e.g. Bourdieu, 1986; Coleman, 1988; Lin, 2001).

Bourdieu developed the labels and thinking around cultural and social capital in the 1970s and 1980s, and his classic view of society and the social world is of 'accumulated history', where an interplay of experiences and attitudes serve to place value on certain signifiers and relations. Often this is formed unconsciously as actors invest or are otherwise inculcated with attributes or resources in order to maintain or advance their position. His analysis has it that all forms of capital exist in relation to economic capital, implying that they are all rooted or given meaning through economic capital. Bourdieu also explains the possibility of 'transubstantiation' or conversion from one form of capital to another, arguing that a variety of resources or capitals are in circulation and are being accumulated and deployed in society beyond the direct economic form. As an example, Forum for the Future (2005) distilled a shortlist of five key forms of capital: 'natural capital', 'social capital', 'human capital', 'manufactured capital' and 'financial capital'.

We can view such capital forms as providing a means of acting or deriving economic or welfare benefit directly or indirectly and sometimes over a long-term 'investment' period. Individuals use one form of capital to gain or build another form of capital. The point that Bourdieu makes about capital, as reflecting accumulated labour, relies on the bearer investing a combination of money, time and energy in developing their resources; each has a 'portfolio of capital'. This view of capital, as a resource that can be directed or used by the bearer, is an important one for planners to understand and there is a clear link between environmental capital and the concern that planning policies have with sustainability (see Chapter 3). We now consider some of the widely used labels or types of 'capital' that are increasingly cited by theorists, policymakers and others.

Types of Capital

There are now many prefixes added to the label of capital used in order to communicate a form or type of capital circulating in society. A wide usage and diversity of such terms have become common currency but they are not always explained or reflected upon. These terms require some unpacking and their application to planning practice explained. We argue that each capital form has an implication for planners and indeed different actions and policies are likely to impact on one or more forms of capital.

Planning theorists have for some years identified and used the capital metaphor to express assemblages and potentials of resources available, or which are required, to produce desired outcomes. Some have also indicated how such relations and resources can act to thwart or distort the best of policy intents. Furthermore, the often implicit aim of much planning policy is to conserve, exploit or otherwise consider the impacts on recognised capital forms. In this latter statement there lies a key issue which can be resolved to a question: do planners fully recognise and understand capital forms in their research and policy? Secondary questions become: what wider effect or impacts do planning policies have on capital forms? We now turn to consider some of the forms of capital beginning with the meta-concept of capital as economic capital.

Capital and economic capital

There are clear overlaps and sometimes different terminologies in use to explain very similar things. Definitions of economic capital are also contested and the term has been broken down by economists with a multiplicity of terms being developed. In the broadest sense economic capital relates to tangible resources that can be used to buy or invest in other categories of capital in the form of money, land and property rights. Bourdieu's shorthand definition of economic capital is as a resource which can be directly and immediately converted into money. It was defined by Marx more narrowly, however, as 'surplus-value', that is, whatever an owner makes in profit from production. Others, largely from a business or economics background, have argued that the term financial capital should be used. This implies a narrowed definition that refers only to money that can be used to invest. Both imply a definite intention to accumulate capital in some way. In all cases the formulations include land and property. This is important given the role that planning policy and regulation plays in influencing land and property values and how such values are calculated (see Chapter 12). This also creates a point of conflict over how to equitably regulate wealth and capital through, for example, restrictions on conversion between capital forms. The control of development rights is the most obvious role that some planning systems play in this respect and allows changes between some of those uses, while requiring that planning permission is sought if other changes are desired. The operation of the system alters the capital value of the land.

Governments are anxious to ensure that national and local economies are not overly constrained by planning regulation and, moreover, that those with economic or financial capital are able to invest to create employment and other means of generating economic growth, including physical development. Of course how such investments impact on other capital forms, resources or 'reservoirs' is the very stuff of the analysis of sustainability and, it follows, of all levels and fields of planning. We now move on to briefly explain some of the range of capital forms commonly discussed.

Cultural capital

Some authors argue that cultural capital plays a structuring role in the accumulation of all other forms of capital. It is seen to be linked to 'human capital' as it is held by the individual, accumulated over time and perishes with the individual. Cultural capital is expressed, according to Bourdieu (1984, 1986), in three ways, as: embodied, objectified and institutionalised cultural capital. These crystallisations affect attitudes, practices and relations between individuals, groups and classes, and serve to distinguish between those divisions. Bourdieu explains each quite clearly; the 'embodied' form of cultural capital refers to lasting or durable attitudes and 'dispositions'. These are learned behaviours that impact on attitudes and opinions of others in some way. They construct the individual and 'place' them somewhere in a constellation of signs, symbols and meanings, along with the other two elements of cultural capital. The second facet is 'objectified' cultural capital, which relates to the ownership or appreciation of objects such as artworks or other prestige goods. The link between the object and the owner and the understanding of firstly, the cost and also the 'meaning' or significance of that item is reflected onto the owner. For example, the owner of a Rolls-Royce may be accorded a certain kudos and gains cultural capital through ownership and association. It may also be that the object, perhaps an artwork, may infer 'taste' or refinement on the part of the owner. Hence visiting art galleries over, say, attendance at football matches has arguably tended to carry more prestige and cultural capital value. Similarly this can be seen in terms of differential values placed on districts or areas and also in terms of ranking certain building designs or types over others. Cultural capital may involve socially recognised academic qualifications, or it may relate to particular habits or inculcated and learned behaviours that signify social status. This form helps explain how investments in property or place may be seen as investments in cultural capital, as well as financial capital; in that some locations are valued more highly not because they are larger in area or because they give rise to direct income streams but that they imbue prestige on the owner. This form of capital also covers the knowledge, skills and education (both formal and informal) that are held by the individual.

The third manifestation is 'institutionalised' cultural capital. This delineates the way in which formal qualifications and awards are recognised and given value in society. Thus, for example, academic degrees and similar attainments provide certain levels of cultural capital that are tradeable. Similarly, status in terms of rank or title can be seen as an expression of institutionalised cultural capital. Membership of professional institutes such as the Royal Town Planning Institute (RTPI), the

Royal Institute of Chartered Surveyors (RICS) or the Royal Institute of British Architects (RIBA) can also provide cultural capital for its members, signifying competency and providing legitimation. The category is rather different, however, from the idea of 'institutional capital' as mentioned below, which is a form of locational or group capital and the aggregate of other available capital resources available to a network.

Human capital

A brief description for this form of capital is all that is required here. Human capital is the term used to explain learned knowledges and skills, abilities and aptitudes held by the individual. It is in effect a net aggregate of cultural, economic and other forms of capital marshalled and invested by that person (or/and their families). Many authors conflate human capital with cultural capital, given that elements such as skills, education and experience are key elements described in accounts of cultural capital (as above and Becker, 1962; Coleman, 2002). Human capital debates have also been supplemented by researchers looking at 'intellectual capital' (e.g. Lundvall and Maskell, 2000). This is particularly seen as important in the knowledge economy, and is relevant for strategic planners seeking to reshape and add value to local or regional economies by up-skilling, attracting investment or new companies to locate to their area. It is also recognised as important in terms of developing capacity and raising understanding of planning and issues mediated by planners. This is most clearly relevant in relation to public or community involvement in planning and local governance.

Social capital

The notion of social capital has been developed and refined for several decades since Jacobs used the label in the early 1960s (Jacobs, 1961). She observed the importance of local social relations present in North American cities and how they were being affected by socio-economic change and urban redevelopment. Many researchers and theorists have since extended the concept of social capital. The concept has been said to be somewhat fuzzy though, given that it covers a wide range of contexts and formulations (Fine, 2001; Lin, 2001) and with numerous dimensions being explored over the past few decades. As a concept, it represents investment in certain types of resources of value in a given society (Lin and Erickson, 2008). Social capital can also be described as the membership of or access to a network of relationships which are the product of individual investment strategies and of cultural behaviours. These result in the generation of group membership and are, consciously or unconsciously, aimed at relations that are useable and beneficial in the short and longer term, as such social capital becomes an individual and shared resource maintained through communication and trust. Coleman more simply puts it that social capital reflects the structure of relations between and among actors (Coleman, 1990). Others emphasise the interdependencies that maintain social capital and some authors talk of social capital acting as the 'glue' which holds society together. Again, the links to networks as discussed in Chapter 4 should be obvious here.

Social capital is not easily exchanged or 'bought', and its development over time and the routinisation of relations of trust are seen as key elements in developing social capital. Bourdieu argues that the accretion of social capital is 'costly' and that it requires continual maintenance. Importantly, several other different elements are also recurrent features in social capital theory. Pretty and Ward (2001) set out four dimensions to social capital. These are:

- trust relations: these are said to lubricate cooperation and facilitate sharing and a culture of friendship;
- presence of exchange and reciprocity, or 'gift' relations: this is where people are happy to enter into arrangements such as lending items, but where some obligation to return favours is also implied;
- presence of rules, norms and sanctions: these are often informal constraints that are observed and enforced by participants or network members;
- 'connectedness', or group or network membership: this may involve meetings or other communications and interactions. This is viewed as an important component of social capital.

In essence the four facets centre around knowing, trusting and becoming enmeshed in moral obligations to a group of people as part of a network of mutual interest.

Researchers studying in what has been termed the 'social capital paradigm' will also be interested in the *quality* of relations and how they are maintained and damaged or broken, as well as *why* those relations are maintained or encouraged. 'Bridging', 'bonding', and 'linking' social capital labels have been devised to demarcate where some relations and efforts to broker social capital through policy have been directed at bolstering relations *within* existing groups or communities and others are directed *between* communities or groups. These may be within groups (i.e. bonding capital), across groups or communities, and in attempting vertical integration – for example in developing trust and cooperation between local government and communities – both of the latter categories are forms of bridging or linking capital (see Putnam, 2000; Selman, 2001).

Physical and other change may destroy social capital as it disrupts the social environment. One of the clearest examples of this came in the 1950s and 1960s, when social relations and the intangible resources built up in community networks were damaged when wholesale urban renewal programmes were embarked on both in the UK and elsewhere around the world. 'Communities' were shifted into new housing and, while many were pleased to be living in improved physical conditions, their ties and local social relations were often broken. One argument has it that such a break played a part in later social unrest, the creation of a community vacuum and in a growth of criminal behaviour (which has been particularly reported in relation to public housing estates in the UK and across Europe). Once relations, traditions and long-term investments are broken or lost, creating or fostering new investment by people in such social relations – in forging social capital – can become more difficult. This becomes a central argument used in trying to ensure that policy and socio-economic change are mitigated when social capital may be damaged. This has also led a wide acceptance of the idea that planning and wider local governance practices should foster 'strong communities' and which harks back to the observations made by Jacobs some 50 years ago (Chapter 14).

Hence part of the role of planners has been to perpetuate or facilitate community in the UK. Despite reservations about the idea and centrality of community in planning (and associated policy rhetoric), there is a direct link between the concept of social capital and that of community as widely or commonly understood. For economic planners it is a common objective to develop social capital as a resource that is useful for local economic development and social cohesion. Indeed, there is a related concern with 'embeddedness' of economic and social activity (see Granovetter, 1985, 2005 and Chapter 18) that links closely to both social and institutional capital. For other planners, including resource planners, the maintenance of social capital can be important in helping to maintain environmental and local institutions. The presence of such relations can also help in terms of community involvement in planning and policy, providing a basis and incentive for civic action.

Environmental capital

Environmental capital can relate to 'natural' features such as landscape and to other bio-resources. This addition to thinking about capital has been developed to reflect and expand on ideas of 'natural' capital and to resources which may be used and managed by human agents. Natural capital is a term used widely, for example by Ekins et al. (2003). It has been broken down into several elements, with authors placing emphasis on different aspects of the environment. The first part of natural or environmental capital recognised in the literature is the productive capacity of land and water. The second is the ability of the environment to absorb waste, with the third being the role of nature and the environment as a basic life-support function; that is, habitats, food chains, the ability to produce oxygen and be protected from harmful radiation. The final element of the concept of natural or environmental capital is the 'amenity' element. This is where the environment is valued by dint, for example, of landscape features, heritage or tranquillity – and it is socially constructed, that is, we place value on these things. In planning terms the recognition and accommodation of environmental capital needs to be made, both in terms of the intrinsic value and necessity of environments for life and the role that nature and the environment plays for health, recreation and human enjoyment.

Environmental capital may not be the direct product of labour per se but it is potentially realised through labour and can be used as a resource by non-'owners' to draw some benefit. It can be seen as both a stock or reservoir, as well as a dividend or flow of benefits. For example, Dartmoor National Park acts as a resource for the tourism industry in using its environmental capital to produce marketing images and providing visitors with opportunities for enjoyment. At the same time, the land provides a basis for food production, it filters water and acts as a carbon sink. This provides a link into another hybrid concept, particularly for those working in countryside and strategic planning. The label 'countryside capital' has been lighted upon to express the elements of the countryside that may be viewed as a stock of resources that can be used and managed for socio-economic benefit. As such the idea draws from social, human, environmental and economic capital forms. The main planning concern is to try to balance the exploitation of such resources with their careful management and to provide a more sustainable rural economy and society.

Conclusion

Essentially capital labels have been developed to explain where and how resources reside and are exchanged. These have been extended recently and newer labels have been coined in order to speak more directly to policymakers and others, with specific or sectional interests and to encapsulate the types of resource circulating and shaping society. Each label divides off an area or type of concern that different actors may have. These aim at focusing attention on the social construction of a particular aspect of resources which reside with people or places and which are given meaning or value in their interaction and exchange. Institutional capital, for example, has been coined as a means of expressing the aggregate of resources available at the level of an institution or local network. However, this appreciation does not necessarily consider how the wider network and the jurisdiction can be seen together as a complex of capitals that are affected and shaped by planning activity. More recognition and attention to such assemblages could provide part of the answer, we think, to more effective and integrated or sustainable spatial planning; and this combination could usefully be expressed or labelled as 'spatial planning capital'.

In brief, the way that planners develop policy and make decisions on behalf of others affects the strategies of capital accumulation and expenditure and can induce, minimise or block certain exchanges or conversions of capital. Planners are also closely involved in resource management and have a key role in shaping, enabling and protecting economic, natural, cultural and human capital or resources. Planners on behalf of society and ostensibly in the 'public interest' act with the state, local stakeholders, politicians and others to effectively valorise certain processes of capital accumulation and types of capital. These values and preferences are reflected in policies and plans which seek to organise, manage and then 'spatialise' capital. This is clearly seen in protected areas, or where development may be actively encouraged in allocated development sites. In these ways planners use and affect different capital resources that shape plans and policies. Knowledge of how and why capitals are formed and exchanged and understanding of the concept(s) themselves are needed as part of the general education of planners. This enables reflection about the resources (or capitals) available and which are 'in play'.

FURTHER READING

This topic area and the combined set of conceptual labels discussed here have been discussed widely. For more specific accounts of capital and planning there are few books or papers, although Healey (1998) and Khakee (2002) have written on institutional capital in planning and Forrest and Kearns (2001) have applied some of the ideas introduced here in an urban context. Key readings on social capital are still Coleman (1988) and see also Lin (2001, 2006). In environmental planning terms, see Ekins et al. (2003) and Selman (2001). Key overviews on social capital are of course found in Bourdieu's work (1984, 1986, 1989) and commentators who have subsequently used, extended, critiqued or applied his ideas, for example Jenkins (1993), Anheier et al. (1995) and Fine (2001). Some planning theorists have usefully explored Bourdieu's ideas too, for example Hillier and Rooksby (2002) and Howe and Langdon (2002).

16 IMPACTS AND EXTERNALITIES

Related terms: public goods; 'Polluter Pays' principle; sustainability; rights; regulation; utilitarianism; freedom; change; environmental impact assessment; mitigation

Introduction

This chapter outlines how planning has developed ways of thinking and sought to apply certain tools and techniques to try to mitigate or avoid so-called 'externality effects' that are derived from human actions. The recognition of various impacts of growth and of associated unplanned development in the nineteenth century gave rise to much of the early thinking and efforts to plan in the UK and internationally, and more specifically the need for some form of planning control (Booth, 2002). The lessons learned in this period of urbanisation and industrialisation produced a need to think about how to mitigate or redirect negative impacts and to arrange land use efficiently with some regard to various quality of life considerations. This thinking provided a foundation for a planning sensibility which was concerned with the environment, public health and efficient land use, and which has since grown into the more multifaceted notion of sustainable development (Chapters 2 and 3).

A development of this early thinking about planning was specifically about the social and environmental impacts of urbanisation, and the early justifications for planning activity which have been extended and updated to fit the challenges of climate change and the carbon reduction agenda of the twenty-first century. The focus of this chapter is specifically on the idea of externalities and the impacts of human activity, and how planning tools have been devised to address or respond to these. Externalities and impacts are key ideas in planning because they are a prime justification for the interventions that planners want to use to shape market processes, or advocate for socially efficient resource use. These ideas highlight the potential role of planning in acting to help optimise decisions about land use and development type, scale and location. The role that planning plays in forcing an examination of wider impacts is also part of a precautionary approach to non-renewable environmental resources and other public goods, which we demonstrate through a short exploration of environmental impact assessment (EIA).

Externalities, Impacts and Planning

Klosterman (1985) outlined four basic functions for planning in the context of the free market critiques made in the 1980s and prior to a widespread take up of the mantra of 'sustainable development'. This fourfold justification involved: protecting the interests of the community; improving the information base for individual and collective decision-making; the protection of the needy in society (and linked to questions of environmental justice); and, lastly, to consider the external effects of individual and group action and respond to the impacts and externalities of (possible) actions. These justifications led to debates about how well planning systems and techniques have performed and whether state intervention does lead to more efficient outcomes. Regardless of such questions, however, the recognition of the importance and control of externalities remains a key issue and a prime justification for planning control and the production of planning research.

Actions are usually purposeful: sometimes they are planned, sometimes more spontaneous. The intention or aim of the action is usually clear but the impacts or effects are not always anticipated, understood, straightforward and therefore recognised as desirable. There is a difficulty in mapping, measuring or understanding the variety of impacts and repercussions of particular actions or changes when examining planning scenarios. Despite the challenges in fully assessing these it is important that planners and others recognise the main or likely impacts of change. Linked to the wider justifications of planning described above is the idea that unregulated or unconstrained individual or group acts can lead to public 'bads'. These are referred to here as negative externalities. In planning terms, such externalities would usually be regarded as being contrary to the wider 'public interest' (Chapter 9) and planning tools are created so that likely negative externalities are avoided, managed or mitigated in some way. One of the intentions of many planning systems is to provide a framework to understand, direct, shape or deflect various impacts. Some impacts can be more easily managed and this may be possible through the reconfiguration of development or by directing certain developments spatially, for example by zoning to ensure that industry is concentrated away from residential areas. Further actions may be taken to absorb an externality effect; one example is the provision of money by developers to public authorities to provide transport infrastructure where new employment, shopping or leisure trips may be generated by a proposed development. It is also true that impacts from development or other actions may be desirable and a role for planning here could be to anticipate and coordinate to maximise such positives, as with the policy of clustering economic activity.

An externality can be said to exist when an individual action affects the 'well-being' of another individual. The broadest definition of externalities sees them as 'spill-over effects'. This implies that externalities go beyond the scope, legitimacy or geographical space where they emanate from or are deemed socially acceptable. Externalities are often seen by economists as 'market imperfections'; they are viewed as the effects of an action which are not fully reflected in market costs and where costs are passed onto other parties (who have little or no ability to avoid the impact). Impacts are in some sense broader and can be extended to mean both intended and unintended outcomes or side-effects that are seen as positive or negative. In some instances, the impacts of development may be calculated to be positive and create jobs or perhaps to improve quality of life through the provision of say a hospital facility. However, the way that impacts and externalities are usually discussed in planning tends to focus, rightly or wrongly, on *negative externalities* or impacts that need to be avoided or otherwise dealt with.

In this discussion, we mention liberal political economy dating from the eighteenth and nineteenth centuries. Here the individual's freedom to act was stressed, but it was also recognised that such freedom to act should not impose a burden on another. Around this time the role of the state and of legitimate state action was being debated in parallel to questions of individual liberty. One of John Stuart Mill's well-known maxims indicates a principle that has been applied widely:

> The only freedom which deserves the name is that of pursuing our own good, in our own way, so long as we do not attempt to deprive others of theirs, or impede their efforts to obtain it. (1859: 16–17)

The principle that individual freedom is to be encouraged is one that has become an important tenet of English law and has been widely adopted around the world. Importantly the appropriate limits to this freedom have been generally agreed to be where the exercise of such freedom does not impinge on others. Where and what constitutes an infringement or what disadvantages another person is, of course, difficult to determine. Indeed, efforts to redress such situations are, by definition, after the fact and may be too late for effective remedy. This brings us back to the role of the state in acting to intervene before (as well as after) such actions, so that the negative impact is prevented, managed or mitigated. This is done as part of attempts to promote equity, efficient resource use and to minimise conflict. For planners this is largely orchestrated in terms of land use and land use activity and in a recognition of adverse impacts being a primary justification for regulation and forward planning activities.

When we consider the actions of individual actors, or particular interests, in spatial terms (and across the social, economic and environmental spheres), there will be many occasions where preferred or intended actions could place burdens on others, or where individual actions over time can act to compound a problem. For example, if everyone in a street were allowed to redevelop their house into a block of apartments there would be a range of incremental impacts, including in terms of water use, traffic, loss of light and amenity. Some of these would need to be borne by the wider public, others would impact directly on neighbours and could not simply be internalised. Sometimes such intentions can be negotiated and arrangements made such that impacts or externalities can be planned and managed. In the UK a series of measures to negotiate around such tensions, and where change is brokered, have been attempted and are discussed below. Furthermore such impacts or externalities may not be always regarded as negative and could lead to net benefits or be linked to strategic goals. In many instances, particularly where development involves employment uses, a mix of positive and negative impacts is often anticipated. In that sense, 'good' planning involves mitigating or ameliorating the negative and maximising or integrating the positives. This is also where, on balance, the wider public interest is calculated (see Chapter 9) and negotiated between parties.

Positive or negative impacts may be deemed acceptable or unacceptable and arguments about what degree of planning control or regulation is appropriate or necessary and who effectively 'pays' or internalises externalities, or where societal preferences constrain others' interests, are ongoing debates in planning and economics (e.g. Alexander, 2001). The production of particular environmental 'bads' such as pollution are an often-cited example; infamously the fumes, smoke and gases that made up the ubiquitous smog that pervaded London and other industrial cities in Britain, did have a considerable influence and acted as a prompt for modern planning control. The impacts or costs of increased pollution would not have usually been borne by occupiers or landowners and they would be detrimental to public goods such as clean air and potable water needed by society at large.

Education and awareness-raising may remedy some problems of inefficient resource use, but it is regulation, such as density ceilings for development and the allocation of particular types of development away from other land uses, that are good examples of planning direction and control. Licensing for particular activities is another method of regulation, including limiting water use. Another option is the use of taxes or fiscal incentives to reduce negative externalities, or to recover some of the costs that arise from 'polluters' or otherwise, to direct actors to adopt other behaviours. An example in this regard would be to incentivise new developments to make use of green technologies using taxation, or to create a 'carbon credits' system (see Sedjo and Marland, 2003). Questions of how societies choose to allocate resources have been subject to prolonged debate, with alternative solutions such as more reliance on market allocation being periodically proposed over traditional planning controls and state regulation (see Evans, 2004).

Externalities and impacts may also be viewed as socially constructed or subjective; that is to say a judgement is often made about whether the externalities provide positive or negative impacts or whether and how to address them or indeed if the magnitude, importance or long-term effects are significant enough to warrant intervention. Sometimes this will be difficult to judge as both elements could be present and, indeed, planning has often been faced with a task of trying to balance these and to justify trade-offs. For example, between creating jobs, or maintaining economic activity, and with concerns over the environment or quality of life considerations playing a part in deliberations.

One area that will always be contested is where different parties argue over aesthetics and the loss of rather nebulous characteristics such as 'amenity' (Chapter 18).

Planning Tools used to Address Impacts

We can see that there are different ways to understand and address externalities and various impacts. Some lie outwith the scope of traditional land use planning, but are part of a wider set of planning and regulatory tools. Now we consider more specifically some of the planning tools that fall into the categories of regulation, taxes and incentives, and awareness-raising. Numerous planning policies and tools have been devised to try to target particular externalities, and plans more generally have attempted to create a context or framework which involves the mitigation, amelioration or avoidance of particular impacts. This is achieved by carefully thinking about the appropriate juxtapositioning of land uses, through the regulation of both land use and the production of spatial plans. The introduction of specific regulations or policies that guide development, or which clearly set standards or thresholds that are designed to limit impacts, is part of this approach. Setting densities for development is an interesting example as it illustrates the different dimensions of externalised impacts. In some circumstances lower densities may be desirable to maintain particular environments, for example in some rural areas this may be true, while in urban areas higher densities may be deemed appropriate to maintain services and ensure viability of high-quality mass transport and respond to policy that seeks more residential development. The second category alluded to above relates to taxes and incentives and how these can be directed towards encouraging and discouraging particular behaviours. Planning agreements and planning conditions in the UK (Chapter 10) carry some of the characteristics of such measures, as they seek to help cushion or organise impacts of development and, while not strictly speaking taxes, they set out requirements that are needed for development to 'perform' in particular ways in order to avoid negative externalities or minimise unwanted impacts. The third category involves education and awareness-raising, and the example of environmental impact assessment (EIA), discussed below, involves an aspect of this. We can also examine impacts in terms of their 'type', as different impacts may be largely defined as affecting the economy, community, social fabric or the environment – although separating these is not so straightforward.

Types of impacts

Impacts or externalities may be tackled in various ways, but are not always foreseen; but they may be broken down following the standard tripartite sustainability dimensions. As such we may consider social impacts as an effect of an activity on the social fabric of the community and on the well-being of individuals and families. While defining this type of impact is not straightforward, it centres on the notion of quality of life and touches on questions of amenity and of public health. A negative social impact might be where a new development builds on the last area of green space within a given locality and no other recreational area is accessible to local residents.

The recognition of, and a method of assessing, environmental impacts is discussed a little more later on, but these are where much attention has been focused in planning over the past 20 years or so. Chapter 3 on sustainability discussed the issue of the environment and how this has featured as a key consideration for planning over time (albeit taking on slightly different forms and carrying different labels as time has passed). One way of thinking about environmental impacts is to consider pollution, carbon emissions or landscape and amenity value that will be produced and possibly lost. Many aids are available to guide planners here. There are checklists, hierarchies and standards against which proposals will be judged. The scientific and the economic costing of environmental benefits or goods are difficult and this area has proved particularly problematic for economists to deal with (Hanley and Barbier, 2009). Cost–benefit analysis, contingent valuation and hedonic pricing methods are just some of the models created to approximate values. Other professionals have sought to create guides and maps, and to model ecosystems to try to aid decision-making where possible environmental impacts are anticipated. The final category here concerns economic impacts, which are important considerations in terms of local and national competitiveness and of the likelihood of job creation or investment. Economic impacts and the possible externalities of economic growth are often where the three types of impacts can conflict or be in tension. The classic argument for development will tend to be couched in terms of the benefits or positive impacts that it will have in economic terms. Again, planning plays a key role in negotiating and coordinating in cases where there are such possible impacts, stepping in to arbitrate between different interests (and sometimes 'getting the blame' when the resulting outcomes affect one or more interests negatively).

The impacts of planning

Many authors, and indeed governments, have been interested in the possible costs or other impacts of planning. This is primarily driven by those placing economic development above other factors, or seeing market processes as the best allocators of resources. Contestations over allocation models and decision-making reflect attempts to ensure that the processes and the outcomes, rather than the principles involved, provide justification for what we do (rather than why we do it). Rydin (2003) discusses the impacts *of* planning as does Evans (2004) and both point to issues of delay and other costs incurred by operating plan-led systems. Such questions and arguments do not convincingly address market differentials and the use of scarce environmental resources (such as agricultural land, or irreplaceable environmental goods), or adequately rebut questions about the need to think long term on the one hand as well as ensure that shorter-term benefits do not damage or contravene the futurity principle mentioned in Chapter 3.

Restricting development in rural or non-urban areas is a good example where a mix of positive and negative impacts have been identified and weighed up since the late nineteenth century and where externalities of other activities can be reviewed and compared. Webster (1998) sets out some interesting arguments over justifications for and against planning and the types of planning tools used. Zoning is one approach used as an attempt to provide a basic framework to organise and anticipate likely impacts and externalities – seeking to intervene lightly but clearly indicating types of uses appropriate in different areas and highlighting the entry costs that may be faced by developers in the forms of taxes (or indeed incentives or benefits).

Measuring and Understanding Impacts in Practice

Using the example of EIA, we show how some anticipated impacts and externalities are managed by the planning system in the UK, and similar requirements are now in force across the EU, with many other countries also opting to adopt variations on EIA. EIA is a widely used tool to check what likely impacts and externalities would follow from a particular proposed development. It is essentially a process to ensure that reflection and examination of the impacts of development – particularly in reference to environmental impacts – are carried out by developers and are checked and considered by planners. In the UK these regulations were first introduced in 1988 and latterly revised and extended. The main purpose of EIA generally is said to be:

> to give the environment its due place in the decision making process by clearly evaluating the environmental consequences of a proposed activity before action is taken. The concept has ramifications in the long run for almost all development activity because sustainable development depends on protecting the natural resources which is the foundation for further development. (Gilpin, 1995: 3)

Others claim that EIA also serves an educational role in informing and highlighting to all those involved what the long-term impacts of developments may be. By doing this, decision-making should be steered away from proposals that are deemed unsustainable. The typical elements of an EIA start with, firstly, identifying the key issues and concerns of interested parties (i.e. a scoping element). Secondly, a screening process is usually undertaken to decide whether an EIA is required and, based on the information collected, the identification and evaluation of alternatives to the proposal. The next element is usually in suggesting mitigating measures to deal with identified impacts and to review proposed actions to prevent or minimise the potential adverse effects of the project. Lastly, an environmental impact statement, which reports the findings of the EIA process, is produced. Thus the EIA process is designed to aid planners in making decisions over development proposals and can act as a way of inviting developers to reflect on their initial ideas. EIA are usually only carried out for larger proposals and where there are likely to be significant impacts. Issues relating to definitions of impacts, and the scope or range of what is considered (including the idea of cumulative impacts), are ongoing features of planning law and practice debates worldwide and, in this sense, tend to mirror debates and definitions applied with respect to sustainability and sustainable development.

Despite the logic and the benefits of such an approach, the application of EIA has been criticised as the assessments may be biased or incomplete, or otherwise 'slow' the planning process. In spite of such criticisms, EIAs remain an important locus of control and contestation over larger-scale development. They show in operation how planners and other interests are required to demonstrate that impacts are known, that measures are in place to mitigate them, or otherwise, to indicate where there is a need to change the scale or type of the development proposed.

Conclusion

The idea of externalities and of impacts provides us with conceptual labels that indicate how actions, and particularly development, produce wider effects. Some of these impacts may or may not be anticipated, and some may be deemed to be positive and may be socially constructed; subjectified, as 'acceptable' or 'unacceptable'. Some proposals may involve impacts that may be considered to be unacceptable in planning or environmental terms and cannot therefore be permitted (e.g. redeveloping a cathedral or building large housing developments in protected areas). In planning terms, a system to manage impacts or absorb them can be developed – some may say to effect compromise in the face of environmental and social concerns. The EIA process, for example, shows how planning has a role in understanding the likely externalities and assisting in managing these, along with other mechanisms, including the operation of private property rights and the legal system. In parallel the use of taxes and incentives and of tools such as planning agreements may be used to facilitate development and to manage and otherwise control impacts. This aspect of planning is potentially very wide, and recent practice in the UK has recognised this and widened it to include a more general 'regulatory impact assessment' (RIA). PSS 5 for England, published in 2010 (and due to be shelved in 2012), requires that local planners undertake a historic EIA to specifically assess likely impacts on historic or heritage environments (DCLG, 2010). The way that the concept and particular impacts are defined are changeable therefore. Policies represent an ongoing part of efforts to ensure more sustainable forms of development and that longer term, acceptable development patterns are delivered.

It should be clear from the above that the task of trying to anticipate and then regulate or steer impacts and externality effects is a burden and a difficult role for planning and planners. There are ongoing debates about the range and complexity of the task and the ability of planners to actually manage this; particularly given the weight of influence against preventing development that may produce some less desirable impacts, or some degree of negative externality. Instead the onus has rested on planners to try to negotiate and mitigate these in such conditions. The process is therefore politicised and can be both resource-intensive and take time: none of these features make for warm interactions between planners and developers. As such the EIA effort to raise awareness of wider impacts is, notionally, a good one, but rather limited. More will need to be done to organise regulatory frameworks and overall planning toolkits to manage impacts if the climate change agenda and wider quality of life questions are to be convincingly answered.

FURTHER READING

There has been much discussion regarding externalities in the economics literature and in debates over planning. Much of this provides basic outlines of what externalities and impacts are and the context of public goods. Rarely are such accounts applied directly to planning, but Evans (2004) does provide a view about the impact *of* planning rather than the impacts that planning seeks to manage, and Rydin (2003) also covers this. Webster (1998) takes a more philosophical approach in examining the evidence and the justification for how we plan and the relationship with negative externalities. The issues concerning the justification for planning have been examined over time by authors such as Polanyi (see also Alexander, 2001; Sternberg, 1993) and Lai (1994) on zoning, alternatives to land use planning and consideration of addressing market failure. Klosterman (1985) explains the wider social and environmental issues raised by negative externalities and the justifications for planning in the face of such impacts. Efforts to try to understand environmental impacts have been devised and centre around processes, such as EIA (see Glasson et al., 2005; Morris and Therivel, 2001; Petts, 1999), while wider views of social and environmental impact are introduced by Barrow (1997). Critique of EIA is developed in Cooper and Sheate (2002). For more on the up-to-date requirements and scope of EIA, see national policy and statutory regulations. In the UK, the government provide guides to procedures for EIA for planners, and which are revised from time to time.

17 COMPETITIVENESS

Related terms: capitalist economic development; globalisation/glocalisation; benchmarking; quality of life; investment; attractiveness; capital; innovation; productivity; networks; economic clusters; governance

Introduction

In planning and economic development terminology, the word 'competitiveness' is often used as a way of describing rivalry and reflecting the relative performance of firms, cities and regions, as well as driving policy and measurement across a range of social and economic indicators. It is also a term that is widely used but one that is not terribly well understood. It has a particular importance for planning activity, given that the competitiveness of regions has become: 'an issue not just of academic interest and debate, but also of increasing policy deliberation and action' (Kitson et al., 2004: 991). Indeed economic competitiveness has become a key concern for many national governments, regional agencies and local authorities and is often cited by a diverse range of public–private partnerships that are involved in restructuring and promoting particular places.

The notion of competitiveness stems from a longstanding interest or focus in economics and economic geography, which concerns the relative performance of economies, and this has been debated widely (see, for example, Begg, 2002; Malecki, 2007). We will not delve too deeply into the specificities of those debates as they have generated and sustained a small industry of themselves. Instead we seek here to underscore the way that planning has a role in influencing, and is simultaneously being influenced by, the concept and rhetorical use of 'competitiveness', particularly for strategic planning. We briefly discuss the development and characteristics of competitiveness, as generally used in public discourse first, before turning to focus on planning.

Competitiveness: What is it?

Competitiveness has a root in economic theory concerning economic competition, trade and the development of modern states, with classical economists such as Ricardo referring to the idea of 'comparative advantage' where trade benefits and strengths reflected differences in resources and assets held by different states. This understanding was used to organise those resources and their trade in assisting national economies. This was originally centred on a concern to ensure the best use of domestic resources at a national scale and how to benefit from imports derived through international trade. This emphasis is countered by business studies theorists who focus more on exports and job creation, while others also point to a wider range of issues that impact on supply and demand for goods and services and the ability of economies to respond to change, such as; trade barriers, tax regimes, exchange rates or workforce skills.

Here we cannot provide all of the details of the different positions and emphases, but echo the hybrid 'strategic-realist' account provided by authors such as Krugman (1996) and adopted by many policy planners. This is where all the elements are seen as important, along with the role of government in helping to support and assist business interests. Furthermore, as global or 'glocal' trade has developed, 'competitiveness' has assumed a greater significance for planners at the scale of regional, urban and local economic development activity (Brenner, 1998; Malecki, 2007; Porter, 2003). This is where we place our focus, as it tends to have the greater impact on planning discourse and practice and also chimes with efforts to relocalise and embed economic activity. We return later to consider the role that planners play in shaping conditions for competitiveness and how, correspondingly, the term is often used to put pressure on planners by various others.

Given the above divergences there are competing definitions of the term, with numerous dimensions presented in the literature, partly because of the different scales where competitiveness is expressed. There is also controversy over what constitutes relevant competitive factors and therefore what and how 'it' is actually being assessed or measured. Furthermore the emphasis placed on different factors and measures varies across academic disciplines and spatial scales, and different policy aims often serve to obfuscate the concept. The UK government view has reflected an emphasis on firms, seeing competitiveness as 'the ability to produce the right goods and services at the right price and at the right time. It means meeting customer needs more efficiently and more effectively than other firms' (Department of Trade and Industry, 1998: 2). While this is pitched at the level of individual businesses the term is also operationalised at larger scales with the generation of economic indicators and benchmarks and business support seen as a key role for government. As such, the planners' role is conceived here to be anticipatory: that is to provide for the needs of industry and the economy in a strategic way to ensure efficient land supply and infrastructure provision. In essence then competitiveness can be seen as the success with which different places (cities, regions, nations) compete with one another and how the economy and available actors and resources (e.g. firms, labour, natural resources and ancillary services) are facilitated and supported and crucially how this milieu provides for the needs of the population.

Competitive Places

The use of competitiveness as a concept here involves questions about *how* places compete and *what* assets they need to compete with, as well as the *spatial dimensions* and relations of competition and productiveness that are present or seen to be required (see; Kitson et al., 2004). We retain as our main focus here 'place competition' and in particular regional and local competitiveness, given its prominence in policy terms and the direct influence this has been having on strategic planning over the past 15 years or so. Storper (1997) provides a simple definition of 'place competitiveness' seeing it as 'the ability of an (urban) economy to attract and maintain firms with stable or rising market shares in an activity, while maintaining or increasing standards of living for those who participate in it' (1997: 20). It also implies therefore inter-place rivalry and this can lead to some distorting or unbalanced outcomes (Kipfer and Keil, 2002), depending on the measures used and policy tools and responses adopted (Begg, 2002). Therefore warnings have been voiced over time about understanding overall impacts and prioritisations of planners, government and industry (i.e. the key actors in the 'governance network'). The concern being that some policies may damage certain resources or assets (or other places) in the process of attempting to exploit and maximise or 'compete' in economic terms.

Some key authors have urged a wider appreciation of factors that influence the competitiveness of places (see Budd and Hirmis, 2004; Deas and Giordano, 2001; DTI, 1998, 2004; ODPM, 2003b, 2004c; Porter, 1992) and, often rooted in institutionalist economic analysis, they have identified how sustainable, or sometimes labelled as 'SMART', growth may create a more 'sustainable competitive advantage' (SCA) and locations where aspects such as social, human and environmental capital are concentrated are seen as important (e.g. Porter, 2003; Chapter 16). Competitiveness is increasingly being seen therefore as a concept to be deployed in terms of, and in association with, sustainable development and the overall performance of places across social and environmental dimensions, as well as economic ones. Porter also argues that to understand competitiveness and to try to ensure that a place is competitive factors such as innovation and institutional capital need to be understood. This brings into sharp relief the practices of firms in trying to control costs and overheads. Porter claims that:

> competitiveness is a function of dynamic progressiveness, innovation, and an ability to change and improve. Using this framework, things that look useful under the old model prove counterproductive. (1992: 40)

This 'old model' is perhaps most pithily expressed by a well-worn quote from Henry Ford: 'Competition is the keen cutting edge of business, always shaving away at costs'. However, this view has been somewhat superseded by attention given to excellence and *innovation* and the wider conditions required for those to flourish (Montgomery, 2007). Oinas and Malecki (2002) reviewed the way that innovation is important in understanding competiveness and productivity, seeing innovation as an important feature in remaking and improving economic performance and productivity. Krugman (1996), a critic of the idea of national competitiveness, argues that the term might as well be a synonym for productivity, a comment that has found some support, notably with Porter (1990, 1998, 2003). The attention that cities and regions pay to innovation and the role of human capital has led to a focus on more micro-level analysis, that is sub-national down to the level of the firm, and specifically concerning 'industrial districts' or 'clusters' (Cumbers and MacKinnon, 2004; Martin and Sunley, 2003). This has led to many areas seeking to focus on the development of hi-tech, bio-tech or other 'intelligent industry' as a means to bolster their competitiveness and regional productivity (e.g. Tomaney and Mawson, 2002).

As a result of the wider appreciation of competitiveness and the range of factors that contribute to it, attention has been extended to consideration of quality of life measures, social capital and 'embeddedness' issues (see Chapter 16; and Granovetter's work, e.g. 1985) and other soft factors or relations; sometimes referred to as 'untraded interdependencies' (Storper, 1997). A high degree of 'institutional thickness' has also been recognised as important, including the ability to intervene and support industries (see Amin and Thrift, 1995; Henry and Pinch, 2001). Although it should be noted that the idea of SCA does not quite employ the same anticipation of sustainability as some would wish (see Chapter 3). What we can say is that the dominant view has it that the ingredients for, and influences on, competitiveness are wide and difficult to measure. However, low taxes, clear and consistent regulation, skilled labour, the presence of support industries and interrelations, as well as a recognition of the impact of other 'factor' costs are often seen as important in maintaining competitiveness.

Systemic competitiveness refers to the network of support and relations that maintain or might engender competitiveness at different scales. Competitiveness has been seen as a product of cooperation between public, private and sometimes the third sector (Lever and Turok, 1999). This rests on the idea that different scales, that is, national, regional and local (along with a range of other factors), play a part *together* in creating the conditions for competitive advantage. In this vein Malecki argues that the stability of the economy is maintained by 'a tissue of supporting, sector-specific, and specialised institutions and targeted policies ... and on governance structures that facilitate problem solving between state and societal actors' (2007: 640). Kitson et al. (2004) suggest that this view, and its associated justification for strategic intervention, is one that is dominant in many governments. Indeed historically the state has perceived one of its main roles to concern altering costs and barriers (e.g. taxes, interest rates, wage bargaining and subsidies), or in other words in adjusting fiscal and employment policy. However, this role also extends to the 'efficient' operation of the planning system and local regimes of governance, as well as the need to allocate necessary resources, such as land, buildings and raw materials needed by industry. Overall, a better understanding of companies and local/regional factors that push and pull firms, or that influence their decision-making, is seen as an important aspect of economic development at all scales and therefore of concern for planning practice.

Regions, cities and competitiveness

The debate over the applicability of the concept at different scales, what the key variables are, and how to measure it has meant it is difficult to pin down competitiveness clearly. As mentioned, the concept has been applied not only to regions and cities (Budd and Hirmis, 2004) but also to local rural economies (Thompson and Ward, 2005) and to incorporate a range of 'hard' and 'soft' factors relating to *productivity*. These can include both supply-side dynamics and demand issues, for example responsiveness to change and the ability to innovate. In planning, the term has gained particular prominence as countries, regions and localities seek to maintain economic position and performance against each other. As such other places, cities, regions and countries become viewed as competitors. Cities and regions compete for inward investment and jobs, for tourist spend or for one-off benefits, such as central government funding streams or for mega-events such as the Olympics. In this sense they are in 'territorial competition' (Lever and Turok, 1999).

Notwithstanding such issues, the UK government has repeatedly underlined the importance of regional policy that acknowledges and supports productivity and competitiveness as part of a strategic planning approach. For example, it has urged that:

> it is vital that [there is] a coordinated approach to the design and implementation of policies designed the raise regions' productivity and growth... it is essential that a comprehensive package of policy instruments be in place. (HM Treasury, 2004: 14)

Markusen (1996) talks of how some places become 'sticky' in that they attract investment and retain employment. City authorities are interested in such ideas as they prompt questions about why and how public policy should be organised to make places more attractive to investors and to particular types of workers. For example, in the new 'knowledge economy' (Begg, 1999, 2002; Malecki, 2007; Porter, 2003) where typically salaries are higher and activity is 'cleaner'.

The comparative performance of the firm and of localities has been an ongoing concern for government and this has also been closely partnered with the idea of measurement of performance through benchmarking by both firms themselves, and by public institutions who are keen to understand how localities and 'clusters' of economic activity are performing comparatively. Dunning et al. (1998) argue that the term competitiveness is most useful when 'benchmarking' the relative performance of economies, and thus it helps to identify areas of the economy which are lagging. Benchmarking is a process which uses a range of indicators to 'map' performance (Malecki, 2007: 639; see also Huggins, 2003; Tewdwr-Jones and Phelps, 2000). Some view such measurement as important in order to identify and learn from others' practice and policies. This presents opportunities to learn by comparing (Arrowsmith et al., 2004; Malecki, 2007) and can then lead to analysis of opportunities, threats and needs, with subsequent development of the most appropriate and viable policy responses (Huggins, 2003).

Understanding and calibrating such information is fraught with difficulty because of the different contexts that global locations and industries operate within. The way in which data are collected and what is being measured also complicate matters (see Arrowsmith et al., 2004; Budd and Hirmis, 2004). Measurement of competitiveness is also hampered by administrative boundaries and the underplaying of the interdependencies and relations that cross local, regional and national boundaries. The picture painted of local economies is therefore often a partial one and the implications of change and therefore the appropriateness of policy is only partly understood (see Chapter 5). A network approach that could follow the dynamics of relations and encourage more joined-up policy across regions may be useful for improving understanding of interdependencies and of public policy potentials. It is desirable that such analyses should have regard to the overall sustainability of policies and how they are congruent with public attitudes and other extant policy streams.

As competitiveness has risen to become a central idea in economic development, associated research has highlighted both the challenges and the opportunities raised by the increased mobility of capital, labour and by global competition. Arguments used by the aviation industry, for example, often involve references to global competitiveness and their role in maintaining the UK in an internationally competitive environment. They claim that airports are a necessity for national and regional economic development. This argument is used as a reason for allowing the growth of airports and flights and for maintaining lenient taxation regimes on fuel and other duties. Heathrow and the British Airports Authority (BAA) is a classic case in point in this respect, lobbying hard to ensure that the planning system is amenable to its 'needs', as this press release from February 2008 in support of a third runway at Heathrow illustrates:

> In an increasingly globalised and competitive economy, in which companies can move operations more freely than ever, Heathrow – and its ability to compete with European and Middle Eastern hubs – is increasingly important. There is no doubt that the UK is falling behind its economic rivals, because Amsterdam, Frankfurt, Madrid, Paris are planning for the future, building new runways and taking jobs and business from the UK. Today, Heathrow is full and it is time for the UK to signal its intention to compete, for jobs and for future prosperity. (BAA, 2008: n.p.).

This form of pressure (and concomitant resistance from environmental and other groups) to allow such growth has led to recent discussions about how infrastructure is planned for, which culminated in the Infrastructure Planning Commission (IPC) being created under the Planning Act 2008 (but then promptly abolished in 2010 as a new government came into power). Such pressure maintains an onus on governments to continually review and monitor the planning system to ensure that it does not damage competitiveness, as discussed below.

Planning and Competitiveness

There are at least two applicable dimensions most relevant to planning in terms of competitiveness. The first relates to the role that planners can play in the facilitation of conditions for competitiveness; that is, the role of planning as an *enabler*. The other stems from the critique of planning as a brake or *obstacle* to growth, productivity and competitiveness. This latter aspect is the basis for an argument used particularly by business interests and the development industry in pointing to planning as preventing an adequate supply of development land, driving up land and property prices generally and as an important factor in hampering industrial competitiveness overall. The latter claims should perhaps be absorbed as part of a healthy and professional self-reflection. However, the basis for such claims should be carefully assessed, as well as acknowledging the positive contribution that policy can play in promoting efficient land use and the anticipation and provision of necessary infrastructure, as part of the overarching concern to ultimately deliver sustainable development outcomes. Furthermore, an understanding of the wider impacts of growth following a competitiveness agenda remains a crucial part of planning deliberations.

The role of strategic and local policy planning lies largely then in rationalising and directing development and co-designing supply-side factors such as government subsidies, grants, incentives and tax breaks alongside necessary inputs such as land. In the UK, central government guidance has been deemed necessary to direct investment into areas where it is needed, or where it may play some other strategically important role. Tewdwr-Jones and Phelps (2000) point towards the way that incentives are offered to inward investors and how these can also act to distort the 'playing field' for investment in other places, and they cite Wales as an example where this has occurred. Despite some unforeseen or unfortunate secondary impacts these tasks are seen as playing a useful, perhaps crucial, part in regional competitiveness policy. Overall planning activity seeks to understand required resources and opportunities for business locations and acts to orchestrate nodes or clusters of related activity to provide scale economies. One historic outcome is the idea of 'new industrial districts' or economic clusters (Deas and Giordano, 2001).

The second aspect to be touched upon relates to the charge that planning presents extra cost burdens on business. These claims are often ill informed and pay little attention to the role that planning plays in attempting to mediate between interests and other socio-environmental priorities (e.g. landscape protection, prevention of urban sprawl). Indeed there is strong evidence that investors and developers overall like a clear and consistent planning system, as it provides a degree of certainty in their own business planning and decision-making (Tewdwr-Jones, 1999). Despite the evidence being mixed and often biased in some way, the idea that planning (particularly local planning and development control) has a negative effect on competitiveness is still a worry for strategic planners and this lobbying affects political perceptions. The operation of the planning system can be a casualty of this type of persistent scapegoating and there is a sensitivity towards criticism of policies that appear to prevent development, or otherwise attempt to constrain firms, industries or economic activity in specific areas. Fainstein (2001) points out that the need to understand the range of dimensions of competitiveness is important if we are to maintain and recall a role for planning in both assisting economic development and in addressing concerns over social cohesion and environmental protection, as well as designing systems that facilitate economic productivity and stable growth.

Conclusion

Competitiveness has been used widely in the past 10–15 years to indicate a need for cities and regions to measure and understand their own economic performance. This has led to a great deal of thinking about providing support for important industrial districts and other actors in local and regional economies. For some this has focused on protecting and supporting existing economic clusters and for others in trying to attract significant long- or short-term investment as part of ongoing restructuring processes (Tewdwr-Jones and Phelps, 2000). The impact of planning on competitiveness is open to some debate but the link to the sustainable development agenda is clear and the need to plan in relation to other priorities and concerns remains. The literature on competitiveness does, in a circuitous way, recognise the need for strong social and environmental conditions and resources, and the role for planners again is to try to balance the competing claims and purposes of different interests. Unfortunately the rather 'single-strand' critiques from some economists and business interests do not, apparently, adequately allow for the holistic and integrated nature or intent of planning. Nor do they embrace much of the understandings of how attractive places are important for a range of social, cultural and environmental reasons – as well as their economic characteristics.

READING AND REFERENCES

The paper by Kitson et al. (2004) is a useful overview of (regional) competitiveness and Krugman's (1996) piece is designed to take a reflective overview of the debate. Malecki (2007) also provides a critical review of the application and analysis of competitiveness at the regional level. The special edition on competitive cities in *Urban Studies* (1999, 36(5–6) still provides a good overview of different attitudes and aspects of place competition, and the edited volume by Begg (2002) includes useful case study and overview material applied to planning and competitiveness. Tewdwr-Jones and Phelps (2000) give an indication of how interregional competition is played out through policy, while Deas and Giordano (2001) deliver a good overview of the concept and the components that might be important when applying it to cities. In terms of policy, it is useful to look at one of the regional spatial strategies which were developed in England during the 2004–10 period and its accompanying economic strategy to see how competitiveness is framed and what was being measured and aimed at in that region. Beyond this, there is a plethora of material available on related topics such as productivity, innovation and clusters.

18 AMENITY

Related terms: pleasantness; sense of place; place identity; character; environment; environmental justice; liveability; quality of life; heritage

Introduction

Concern for the environment and environmental quality was an early preoccupation for urban and regional planning. In the UK this has roots located in attempts to deal with the ill effects or negative externalities of industrialisation and urbanisation processes that became acute in the nineteenth century (Chapter 16). The experience of such environmental conditions highlighted the need for clean environments and places that were liveable and more pleasant. This harked back to the pastoralism that many had experienced, or held as an ideal for their own lived environment (see Bunce, 1994; Urry, 1995).

The idea of amenity in planning stems from the early idea that characteristics of place were both valued by those living there and that these social spaces and physical environments were being compromised for industrial or other economic purposes. This concept is closely connected to place and sense of place discussed in Chapter 13 and concerned with ensuring sustainable and pleasing environments for people to live and work in. It was perceived that by means of 'good planning' societies could in some way improve or maintain the lived environment and act in the public interest to secure amenity (see Chapter 9). The emphasis on preserving amenity reflects something more, however, than simply creating healthy lived environments; it highlights the influence of the conservation movement in Britain and an associated interest in protecting open space and the wider countryside. Such sentiments are found to a greater or lesser degree globally and may not only be operationalised through planning systems.

With this background amenity is often cited as a justification for refusing planning permission in the UK, or otherwise for intervention and discussion over the design qualities of developments and public spaces. There are policies and guidance at both the national level and at the local level in UK planning that influence this and amenity can be used as a 'material consideration' in reaching decisions over development proposals (Cullingworth and Nadin, 2006; Duxbury, 2009). There are also numerous designations and restrictive policies in place which are, ostensibly, designed to preserve amenity: even if sometimes that term is not explicitly used. This situation has not helped to mollify critics of planning as amenity remains a question largely of subjective judgement and is often mobilised against new development.

Definitions and applications of amenity

Amenity considerations are found or often lie implicitly in much urban and rural policy decisions. The term is invoked commonly in local decisions and sometimes as a shorthand or proxy for the quality of a place. Cullingworth and Nadin state that 'amenity is one of the key concepts in British town and country planning but nowhere is it defined in legislation' (2006: 164). The term amenity centres on some subjective quality of being pleasing or agreeable in terms of the lived environment. In the case of the UK this, in large part, has Victorian advocates to thank for its long-lived centrality in planning thought and practice. Amenity is a cross-cutting idea stemming from a concern with the qualitative appreciations of place derived from a complex of factors that, together, form some agreed positive attributes of place. The way that amenity is deployed may centre on a claim to assume the support or wishes of either a majority of local opinion, or an imposed view of what should constitute local amenity or 'pleasantness'. More cynically still, it is possible that this equates with the maintenance of self-interest and protection of property value, as much as reflecting a wider or shared community interest position.

Amenity may also be a consideration in terms of how *change* may impact on the social vibrancy or make-up of an area. It is therefore a concept that is intrinsically both socially constructed and interleaved with the relationship between people and place. In other words, the dynamic between land uses, semiotics of place and the ideas, experiences and understandings of the people experiencing a given place. As alluded to already, it is an idea that has often been appropriated by elite groups maintaining a belief that particular features and built forms are intrinsically more or less desirable. It has also been used by economists in trying to understand land values and locational decision-making, and often it is conflated with terms such as place utility in this regard (see Bartik and Smith, 1987). This refers, however, more to the usefulness or utility of an area and its attributes, rather than arguments about the intrinsic value or 'character' which many amenity campaigners aim to protect.

It has been asserted that the maintenance and improvement of amenity should be a central aim of planning. Indeed the term is found in the first Planning Acts in the UK, but it is a rather ill-defined concept. Smith complains that 'amenity is easier to recognise than to define' (1974: 2) and as Cullingworth and Nadin (2006) indicated above it is unclear how amenity is reflected in law. However, it is still widely used in planning practice, indicating how many accept the term without giving greater thought either to its meaning, its constitutive elements, or how planning activity and environmental change more generally affects public or civic amenity.

Early efforts to plan towns and cities placed the idea of enhancing and preserving public amenity as a core aim and this was influenced by early planning thinkers such as Ebeneezer Howard, Raymond Unwin, Lewis Mumford and the Garden City movement (see Parsons and Shuyler, 2003; Ward, 1992). Cullingworth and Nadin quote a proponent of the Housing, Town Planning Etc. Bill 1909, which became the first modern-day planning Act in the UK. This identifies the underlying thinking and understanding of planning as enhancing public amenity at that time:

> The object of the Bill is to provide a domestic condition for the people in which their physical health, their morals, their character and their whole social condition can be improved by what we hope to secure in this Bill. The Bill aims in broad outline at, and hopes to secure, the home healthy, the house beautiful, the town pleasant, the city dignified and the suburb salubrious. (2006: 16)

These sentiments, expressed perhaps a little pompously, stemmed primarily from a concern for the health and well-being of city dwellers faced with poor environmental conditions and a physical environment that was seen as increasingly dehumanising. This was coupled with a concern for the protection and provision of open space and recreational opportunities that the Victorian preservationists had championed (Rydin, 2003; Smith, 1974). The quote also indicates how amenity was identified at different scales: for the individual, the dwelling, the neighbourhood and upwards to the level of the town and city (see also Booth, 2002, for an alternative account of the thinking behind the early Planning Acts).

Early planners were concerned then with improving, firstly, environmental conditions and, secondly, the physical appearance of places. These concerns in terms of physical conditions were the main constitutive elements of amenity, or the lack of it, at this time, and were joined by a third element – the preservation of the character of place, particularly in terms of the built heritage. The radical early planning pioneers emphasised a key goal for planning as the preservation of existing amenity spaces and green spaces. It is worth noting that these motives have been incorporated in some way in the more recent idea of 'green infrastructure' (see Kambites and Owen, 2006), which has been developed to encapsulate open space and other semi-natural features, areas and routes.

The term amenity, stemming from this heritage, reflects a conflation of the ideas of a subjective 'pleasantness' and the observable physical and environmental qualities of place. This is quite different from the idea of the term 'amenities', which tends towards the functional convenience and location of facilities, different services, land uses and their interrelation. Although the presence of such amenities can contribute to the overall sense of place and to locational desirability (and therefore form part of an aggregation of wider place amenity) the meaning is much narrower. Another example of the scope and application of the term amenity is in relation to advertising controls through the old Planning Policy Guidance note 19 (DoE, 1992) in England, which stated that amenity includes: 'the general characteristics of the locality, including the presence of any feature of historic, architectural, cultural or similar interest' (para.11). This also counterpoises the rather different meaning associated with amenities.

The experience or makeup of what constitutes amenity may be presented as a result of numerous tangible and non-tangible elements that are affected by different processes of change and restraint. In this sense the consideration of amenity shares a conceptual base with more recently developed notions of place (see Chapter 13; Hague and Jenkins, 2005; Hubbard et al., 2004; Relph, 1976; Urry, 1995). The discussion of habitus in Bourdieu's work also comes close to expressing the complex interrelations and features that are experienced and valued by people 'in place' (see Hillier and Rooksby, 2002; Jenkins, 1993). This includes meanings, memories and feelings linked to spaces as places (see also Crang, 1998). This conceptual extension also problematises normative definitions of amenity – given that different groups and individuals are likely to understand and value different aspects of place – and this also serves to highlight how amenity in planning policy and decision-making is presented and mobilised by different interests. Amenity becomes a reification of dominant ideas of what constitutes the appropriate or the valued in particular contexts (see for example; Andrews, 2001; Bartik and Smith, 1987). More often than not such understandings and perceptions are implicitly reflected in plans, policies and designations, as well as in micro-level decisions (e.g. refusals of planning permission). Such decisions can appear rather petty-minded, and are often criticised as an unnecessary fetter on development and individual freedom.

Over time the term amenity has been displaced in the UK planning lexicon, but it is still referred to in many planning application decisions and it can be deemed 'material' in planning cases (see Duxbury, 2009). Similar and overlapping terms, such as place character, liveability, place identity and sense of place which emanate largely from the urban design literature (e.g. Lynch, 1960, 1981; Punter and Carmona, 1997), have become synonymous with amenity. Wider notions of sustainability (Chapter 3) and of (spatial) environmental justice (see Haughton, 1999; Lake, 1996) have also partly supplanted the focus on amenity, although the central concern for quality of life and creating a pleasant lived environment is still apparent in planning. If one looks closely, 'amenity' is very much ensconced within the challenge and elements of sustainable development and in ensuring quality of life as primary goals for planning policy.

How amenity is measured or objectified is another issue and is a challenge that has spawned a number of approaches: for example, the creation of local character assessments (discussed below) and associated descriptive accounts of the components that are deemed important or appropriate to maintain as being constitutive of amenity. These approaches have tended towards an 'historical type' frame of reference that focuses on heritage and an underpinning preservationism. This conceptualisation was more formally captured by the provisions of the Civic Amenities Act 1967 in the UK and specifically in the creation of Conservation Areas (CAs), as discussed later. However, this view of amenity is still a rather limited one focusing largely on physical appearance, ecology and historic appraisal, rather than the motives of many of the Victorian reformers in seeking to ensure healthy and socially efficient urban living. Instead the term has been somewhat captured by the preservationist lobby. For example in the definition highlighted by the CPRE amenity is described as:

> The pleasant or normally satisfactory aspects of a location which contribute to its overall character and the enjoyment of residents or visitors. The Minister of Town and Country Planning in 1951 stated that: 'anything ugly, dirty, noisy, crowded or uncomfortable may injure the interests of amenity'. Amenity is often a material consideration in planning decisions. (2011: np).

As explained, the assessment of amenity is a subjective one, regardless of any implicit conceptual breadth. This leaves the concept open to capture or narrow application by powerful interests. There has been a long-running debate about the role of planners and preservation interests in insisting on historical pastiche and otherwise 'interfering' in development processes. Not long after the passage of the Civic Amenities Act, Reade (1969) argued that the pursuit of amenity, in the way implied in that legislation carried a danger of it being followed as an end in itself. Particularly in terms of limiting the physical appearance of the built and natural environment and potentially doing so to the detriment of other factors (such as community cohesion, the financial viability of businesses or development schemes and in constraining creativity and new architecture). Without a clear conception and strategy in terms of the social and economic functionality of place amenity, there is a danger that preservationism in the name of amenity could lead to an unbalanced approach to place making where the physical fabric is protected but a loss of social or economic vibrancy may follow.

Dealing with and balancing such issues should be a stock in trade for planners but in practice this can be difficult. Such considerations led to the idea of selectively protecting certain areas or individual buildings by using listed building status and CA designations (see Mynors, 2006). Pahl (1970) highlighted how planners are often confronted with symptoms rather than causes and moreover how attempts to address one symptom affecting amenity can lead to another problem. This does then allude to the multifaceted nature of amenity that concerns not only the environmental and physical qualities of a place, but also how those features are valued and recognised by the community itself. Furthermore, the complex interaction between uses, people and physical environment (Punter and Carmona, 1997) is difficult to replicate or replace and the emergent and evolving nature of 'sense of place' should not be underestimated.

Planning and Amenity in Practice

Our concern with sense of place (Chapter 13) and the importance of heritage in cultural and economic terms also means that measuring, regulating and shaping amenity remains a key concern for planning, and the term amenity is frequently invoked in planning decisions. There are different ways that amenity is typically deployed or understood in planning practice. Two examples taken from planning refusal decisions given by English planning authorities show this. The first quote demonstrates the wider appreciation of amenity as local quality or character, while the second is more focused on the 'amenities' of adjacent occupiers:

> the anticipated high levels of occupation which would lead to an increase in on-street parking in the vicinity and to the detriment of local *amenity* and highway safety contrary local plan policies. (Extract from Derwentside District Council 2008, emphasis added)

> the development by virtue of its size, height, design, massing, window positions and relationship with the adjacent dwellings constitutes an un-neighbourly form of development which would cause overlooking, loss of light and have an overbearing effect on the adjacent properties to the detriment of the *amenities* of the occupiers. (Extract from Rugby Borough Council 2007, emphasis added)

The typical rationale is either that the proposed development would adversely affect the amenity of a particularly place/location, or that it would adversely affect an individual's amenity. Rarely is this unpacked or explored deeply, although some planning authorities have specific policies on local amenity. For example, Carlisle District Council, Cumbria, has a specific policy on residential amenity (policy H2) in its local plan (2001–2016; see: http://www.carlisle.gov.uk/carlislecc/local-plan/helpfr.html). This attempts to clarify how amenity considerations could affect decisions in that area (see also Mynors, 2006).

Amenity is an entrenched idea that is present in planning thought, even if not explicit in policy directly designed for the case in hand. In considering amenity, planners will typically include such factors as impact on views, light, noise and visual intrusion in their decision-making, as well as the overall design and impact of new development relative to the pre-existing built environment. However, this can mean that amenity protection can become a tool in preventing any change, perhaps particularly in some rural areas or where anti-development sentiments are strong. The term was included explicitly in the 2004 version of PPS7, the national planning policy statement for 'Sustainable Development in Rural Areas' which says that planners should: 'have regard to amenity of any nearby residents or other rural businesses that may be adversely affected by new types of on-farm development' (ODPM, 2004e: para. 31). Again the measurement or constitutive factors of such amenity are left to the discretion of the planning authority. Several examples of how amenity is expressed are set out below to show attempts to define and describe amenity in planning and policy.

Landscape character assessment: essentialising amenity?

An example of an attempt to break down and explain components of amenity is found in the environmental planning context and the landscape character assessment (LCA) process developed by the Countryside Agency and Scottish Natural Heritage in the UK. LCA was intended to set out the 'essential characteristics' of place and to assist planners and others involved in land management to reach decisions over their practices, funding allocations, other regulatory processes (such as plan preparation) and in negotiating and determining planning applications. Seven criteria or elements were identified and idealised for inclusion and reflection in these LCAs. These are: landform, soils and geology, land cover and habitats, cultural landscape and archaeology, built environment, heritage, and tranquillity. The mapping process involved in LCAs has led to the creation of 159 'character areas' and a 'joint character area' map for England. LCAs do not involve the social dimension, however, and were intended primarily for rural contexts with a focus on landscape as the name suggests. The main critical charge made against LCAs is that they encourage the idealisation of place and reflect expert knowledges, rather than incorporating wider understandings or values.

Amenity and Conservation Areas

The 1967 Civic Amenities Act enabled the creation of CAs, which are designed to preserve local amenity in terms of a particular historical and architectural heritage. CAs also act to limit the types and degree of variation possible in terms of architectural detailing and land use change in those areas. The chapter on designations (Chapter 8) also makes mention of CAs as the main tangible reflection of planning's ongoing concern with civic amenity and the concern for places that have particular architectural features and traditions that are deemed important. A network of streets which have many Georgian period townhouses, for example, and which are largely unaltered and retain historic value, may attract designation as a CA given that they represent a particular style and period deemed to be important culturally. The powers available to planners using CAs include the ability to restrict otherwise permitted development and insist on the use of specific materials and finishes and to prevent alterations that are not deemed to be in keeping with the purpose of the CA designation. Each local authority in England maintains area-based CAs and each will attract specific policy criteria and stipulations about change. Each planning authority should provide access to information about CAs and the special conditions that apply for residents and owners of property in those areas.

Village design statements

Village design statements are a product of a form of community-led planning with a central emphasis on considering the local amenity and the design aspects of small settlements or neighbourhoods. These are in some sense informal design guides (see the RUDI link given at the end of Chapter 13). Crucially they are put together by the local community in partnership with others, including local planners and they are usually non-statutory documents (although some have been incorporated as supplementary planning guidance in the past, see Owen, 1998, 2002; Tiesdell and Adams, 2011). They are intended to inform developers about the acceptable treatments for new buildings and refurbishment or improvement of existing buildings, based again on the existing styles and other design features found in the immediate area. They are similar in some ways therefore to the aims of CAs but are not statutory or necessarily focused on a particular style, historic character or period. The connection to activities such as parish or community-led planning (see Parker, 2008; Parker and Murray, 2012) with their more holistic approach towards examining and responding to community needs and aspirations brings into focus the emphasis on local needs in social, environmental and economic terms and the wider conceptualisation of amenity discussed here. It should be stressed that where amenity is only deployed in terms of visual character and the local built environment, including, for example, dealing with minutiae such as trees and their role in place amenity, it can actually become socially regressive – acting to maintain place exclusivity when such issues begin to prevent other necessary changes from taking place or in taking up professional planners' valuable and scarce time.

Conclusion

Amenity as a label has been used by planners for over a century as a justification for guiding and shaping development. Amenity is a rather nebulous idea though with numerous components that have been included or omitted and it reflects a compression of various ideas and values. As a result it is difficult to provide a comprehensive definition and it is problematic to pin the concept down, partly because it is a composite of objective, subjective, experienced and mappable features and feelings about place. Many of these are related to the value that individuals or communities put on their areas or features of their lived environment.

The term has been often appropriated and narrowly applied in planning in respect to the physical appearance of places, although when cited in planning decisions a range of considerations may be implicated. Despite the encroachment of other synonyms or competing labels such as 'liveability' and 'sense of place', the term still has currency in practice. A reflection on the wider historical basis of amenity and the aims of the early planners indicates that the concept of amenity reflects part of an overall concern about accessible, liveable and sustainable places. The label refers to the overall qualities of a neighbourhood, town or city (and it would seem particularly in suburbs or villages). When set in this context, amenity is entangled in a substantive aim of planning for sustainable development, rather than simply an idea to be used for pursuing a narrow preservationist objective or a 'steady state' to be defended.

FURTHER READING

The text by Smith (1974) still retains its relevance here and has as its main focus the role and history of amenity in planning. Delafons (1997) provides a nice account of the issues and struggles over preservation and amenity in the UK. Many of the standard planning textbooks briefly touch on amenity (see, for example, Cullingworth and Nadin 2006, Rydin, 2003). The urban design literature includes considerations of amenity (for example Roberts and Greed, 2001) and more detail about how CAs are managed can be found on the Civic Trust web pages. In order to view examples of the LCA characterisations and accompanying descriptions see http://www.countryside.gov.uk/LAR/Landscape/CC/jca.asp. Interesting reading on the development of amenity and early planning thought is found in Ward (1992) and of course the original *Tomorrow: A Peaceful Path to Real Reform*, written by Ebenezer Howard in 1898, and republished by the Town and Country Planning Association (TCPA) in 2003 (Howard et al., 2003), is striking in terms of its continuing influence on planning and planners at all levels. A browse of the regional economics literature will also be enlightening in order to see how amenity is being measured and assessed in relation to locational decision-making and relative land and property value – see Andrews (2001) and Bartik and Smith (1987) for an introduction to this viewpoint. It is important to locate and understand where and how amenity (and its synonyms) are used in policy and practice. Prevailing national and local policies and law sources can also help in seeing how the term has been used and shaped by policy and case law over time: see for example, Duxbury (2009) in terms of English law and practice and Mynors (2006) more specifically on listed buildings and CAs. It is also worth exploring what groups are deemed to be 'amenity societies' in legal terms and how they interact with the UK planning systems.

19 DEVELOPMENT

Related terms: actors; interests; land; change of use; property; regeneration; economic development; community development; sustainable development; networks, externalities; impacts

Introduction

The concept of development is at its broadest a wide-ranging term and there are a number of ways in which planning is concerned with development. The term development also carries a number of different meanings for different actors involved in planning. Development may be both an outcome and mechanism or process for planning and the term may be directed at physical, community, economic and environmental processes and outcomes. As such there is considerable linkage to be highlighted with a number of other key concepts included here which provide related insights to those covered here.

In one sense the concept of development is as all-embracing as the concept of planning as discussed in Chapter 2 and for similar reasons, given that planning is concerned with managing change, much of which involves land and property development. The concept of development can be revealed to mean different things in planning practices and invokes not only the idea of physical change, but also the management of economic activity and wider processes that involve community interaction and support. Each of these three forms highlight the importance of the label for planners and others and the need for it to be unpacked and explained. Moreover these emphases or applications of the term development overlap or reinforce with each other, making discussion of the different interpretations even more necessary.

Definitions of Development

A number of related fields and activities have adopted the term 'development' where their actions and concerns involve or impact on the quality of life of local communities and where interventions are designed to 'improve' social, economic and environmental conditions experienced by those communities. These usages all imply that development involves change and moreover should facilitate some kind of positive change.

The primary focus of planning on matters of land use has tended to restrict the scope and reach of formal planning in orchestrating wider socio-economic development, although in practice the operation of planning systems has wider repercussions across all spheres and may be used to effect wider change. In the UK efforts to extend and join up narrower planning regulation with other concerns and activities has been termed 'spatial planning' with its aspirations towards a more holistic approach to socio-economic change (see Chapters 2 and 3 and Nadin, 2007). However, it is the decision-making over *physical* development or property development and land use regulation that is the primary role and outcome of much planning activity internationally. Yet the more holistic approach inferred by spatial planning emphasises the way that property development is a component of wider regeneration and of a more integrated planning with its aim at achieving sustainable development (Chapter 3). Other elements relevant here include community development and the protection or enhancement of cultural life, including concern with less tangible features or benefits.

The definitions of development that focus on physical change or property development are useful start points here, with Wilkinson and Reed stating that:

> At its most simple, property development can be likened to any other industrial production process that involves the combination of various inputs in order to achieve an output or product. In the case of property development, the product is a change of land use and/or new or altered building in a process that combines land, labour, materials and finance. (2008: 2)

This definition captures the economic and particularly the property-led or physical perspective that tends to pervade the planning and real estate literature (Adams, 2001; Guy and Henneberry, 2000; Wilkinson and Reed, 2008). This conceptualisation reflects a dominant concern with land use change but does not capture other significant dimensions that are relevant to those involved in the production and consumption of the built environment. In this regard, a broader view is reflected in the wider literature on development and regeneration, for example with those interested in the economic, social and environmental conditions that shape societies. As such the ideas bound up with the term 'development' may all be central concerns for planners of different types or in different places.

The wider perspective, at its heart, signifies an orientation towards physical development as a means to an end rather than an end of itself. Therefore, some actors, plans and strategies focus on a more holistic appreciation of the term, or will choose to emphasise less tangible forms of 'development' as being overriding objectives that need to be pursued. Such strategies and actions sometimes encapsulate or couch development in terms of 'sustainable development' or may invoke terms such as 'community development' to reflect an integrated approach or an approach that seeks to develop social capital (see Chapter 15) or to otherwise 'improve' the physical conditions that populations experience. These reflect the early drivers of urban planning in seeking to create a better lived environment.

The label of sustainable development aims to reinforce a more holistic conception of development. There is an emphasis on the integration of the economic perspective with the environmental and social dimensions and it carries the twin ideas of providing for *need* within certain *limits*. This infers that societies must address these issues simultaneously. Such a goal consequently presents a challenge for planners, given the range and complexity that such principles convey and the limited powers and resources that tend to be at their disposal.

Interestingly enough, this new holism, drawn from the world of practice, can be combined with similar perspectives arising from academia, where postmodern or more fluid viewpoints have sought to resolve the conceptual tensions between the economic, cultural and environmental dimensions of the (property) development process, and of the needs and impacts of physical development. Thus, Guy and Henneberry argue for a need:

> to develop an understanding of property development processes which combines sensitivity to the economic and social framing of development strategies with a fine-grain treatment of the locally contingent social responses of property actors ... this interrelationship between culture and capital may provide the key to understanding urban development processes. (2000: 2399)

This demonstrates the potential complexity of overlapping ideas of development for planners and others involved in land and property development. It simultaneously highlights how there are many different roles and priorities for actors impacting on a wider spatial planning. Actors are likely to strategically adopt a narrower or wider view of development depending on their interest or goals. In practice therefore, a series of rather narrow and overlapping conceptualisations of development operate or are maintained by communities of practice or 'epistemic communities' (see Amin and Roberts, 2008). Given this situation, the label of development is both an elastic and a pervasive term.

There are also resource constraints, knowledge demands and political factors that both shape development outcomes and which are routinely faced by planners and other actors. These constrain and facilitate *particular* forms and outcomes of planning and development. The legal definition of development used in UK planning centres on a legalistic conception of physical change and of the construction of new forms in the environment, or of practical changes of land use. The legal definition of development has also been claimed to be the 'lynch pin of the system' (Duxbury, 2009: 131). The UK (and more specifically English) planning law for example, clearly demarcates what constitutes 'development' in these narrower legalistic terms:

> The carrying out of building, engineering, mining and other operations, in, on, over or under land, or the making of a material change in the use of any buildings or land. (HMSO 1990, Section 55)

Given our concern to provide an understanding of the multifaceted concept of development, this definition may not really take our understanding across all the terrain we wish to cover. However, this raises awareness of the more limited frame within which (traditional land use) planning has sought to regulate physical development and changes of land use. This has hinged around the granting of planning permission (or elsewhere zoning permits) and this process, therefore, channels planning activity towards a relatively limited set of concerns and legal 'tests' or in some instances market viability considerations.

The widening of the scope of planning towards 'spatial planning' has set up a series of tensions between the legal ambit (focused on 'directing' development) and planning as an activity that seeks to orchestrate a much wider set of actors, resources, conditions and interrelated policy concerns (e.g. climate change, health, and crime and safety) and which may be read as development as 'improvement' (Haughton et al., 2009). In essence then, a widened conceptualisation of development overlaps considerably with what actually constitutes the central and early concerns of planning itself (as discussed in Chapter 2).

Understanding and theorisation of a broader view of development has re-emerged, partly due to the establishment of sustainable development as the primary goal of planning and also in the light of the spatial planning approach promoted in England between 2004 and 2010, emphasising the interlinkages and interrelations between different actions, places and traditional policy 'silos'. In this spirit we now turn to examine the various perspectives and elements of the broader concept of development and their implications for planning practice.

Development and Sustainable Places

Given the pre-eminence of the consideration of the physical dimension of the development process in planning, we start our deconstruction of the concept of development from this perspective and then expand our analysis. The property development process has a number of key components that have been explored over time. There are five components or elements of (physical) development to examine: 'process', 'scales', 'actors/relations', 'forces' and, 'forms' of development. This breakdown reveals the complexity of development processes and outcomes and the linkages to community and economic development practices.

Development as a process

'Process' has been emphasized here in order to convey the dynamic nature in which physical, environmental and economic development occurs. This is central to the production and consumption of the built and natural environment over time. In simple terms the process can be presented as a series of 'stages' through which property development evolves. This starts with initial investment decision-making (i.e. allocating between property and other 'asset classes'), through site selection, financial appraisal, planning and construction, to building use and reuse. This process is intrinsically connected to wider changes in social and economic priorities and attempts to structure the urban and rural environment. These are said to reflect the priorities and aspirations of societies and interest groups as orchestrated through various regulatory policies and policy tools.

Much of the literature discussing property development assumes a linear process and relative stability of actors and conditions. Numerous models and types of conceptualisation have been created, emphasising actors, structures, events and their sequence. However, it is a contested and often 'unruly' process that leads to different outcomes. The Ratcliffe model (Figure 19.1) for example does not account for other relations and variables. While this portrays a reasonably accurate description of typical actors and highlights a simple linear process, it is a rather inward looking model. There are numerous other factors and influences that shape the behaviours and responses involved, as well as affecting the viability of development or different 'forms' of development involved.

Other conceptualizations of development relevant to planning stress the process element, as with *community development*, where the methods used and iterative nature of community development require that development efforts are ongoing and will need to be maintained to support, inform and 'develop' communities. This is often by encouraging self-help and the maintenance of wider quasi-public goods, such as respect for property, mutual support and relations of trust and regard. Organisations such as the community development trusts (CDTs) in the UK have been involved in work that assists communities in working together: 'all communities will become places where people have a sense of common ownership and pride, places where everyone is allowed to achieve their potential, and places which are socially, economically and environmentally prosperous' (Development Trusts Association (DTA), 2009: n.p.). Such organisations act to support local people to come together and to enhance quality of life. These may involve setting up local structures and organisations, through to orchestrating events and opportunities for interaction. An interesting organisation, which has become an international operation and sits between sectors playing a part in both community and physical regeneration, is Groundwork. This organisation has adopted a project-by-project approach that represents a hybrid of 'community development' with forms of physical development (Parker and Murayama, 2005) and highlights how community groups are stakeholders, sources of information and consumers of physical development.

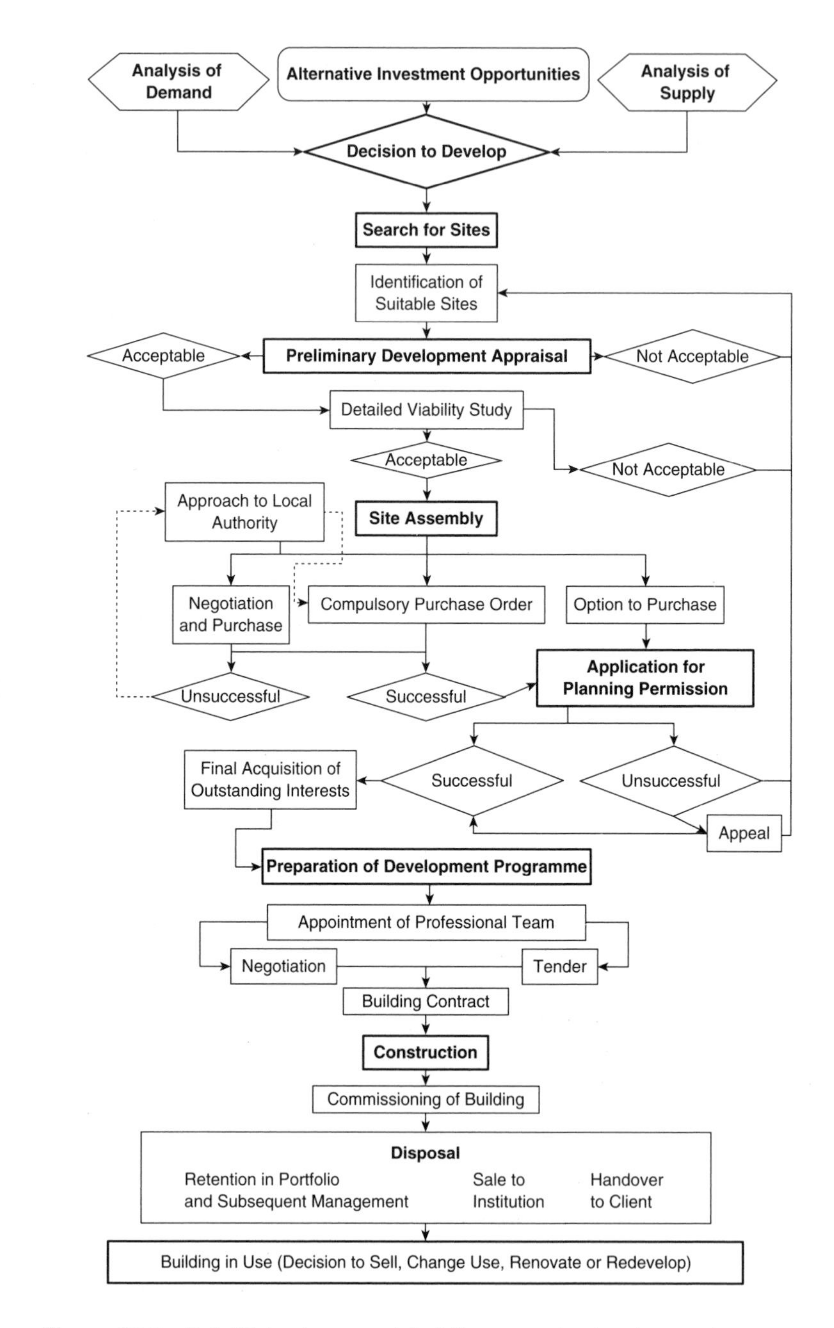

Figure 19.1 Ratcliffe's stage model of the property development process.

Source: based on Ratcliffe (1978).

Relations and actors in development

As already intimated, numerous actors are involved in, and influence, development. Some are easily identifiable and their role or influence quite clearly traceable: for example landowners, planning authorities, developers and investment funds and community groups. There are often other actors and sets of relations that are less obvious, or their role less easily predicted or modelled. Some researchers have seen property developers as the 'orchestrator' of development, or have prioritised the local state as leading, but this does not really encompass the rich tapestry of actor-relations. There is, of course, the possibility of listing or labelling key actors involved in development, however, this approach only takes us so far. There are two points to be highlighted here: firstly, that the list may not be exhaustive or comprehensive, and secondly, that the behaviours and characteristics of the key actors are likely to change from time to time and place to place for a wide range of reasons. This critique stresses that the roles and agency of different actors may differ from development to development. Fisher and Collins (1999) produced an actor-event sequence model (Figure 19.2) which stresses the structural factors, the actors, the site and the events or stages of the development process. To some degree this acknowledges the structural forces that will tend to shape development.

The second component to be expanded upon here is the relational dimension of the development process. Considerable attention has been given to the role of different actors and partnerships in taking development forward, but less on the relations *between* actors. Development is viewed as being permeated and structured by differential access to resources and other dimensions of power. This is inescapably part of the social construction of development discourses, projects and outcomes. As such, information, knowledge and power are key factors in determining the way that actors behave or their relations are shaped and which act to critically shape development outcomes. Awareness of such resources and interactions becomes important when engaging with physical or other efforts to develop built and natural environments.

Forces of development

The third component that warrants discussion is the multi-dimensional (i.e. economic, social, technical, legal, environmental and cultural) aspects of the development process. Operationally these are forces that shape

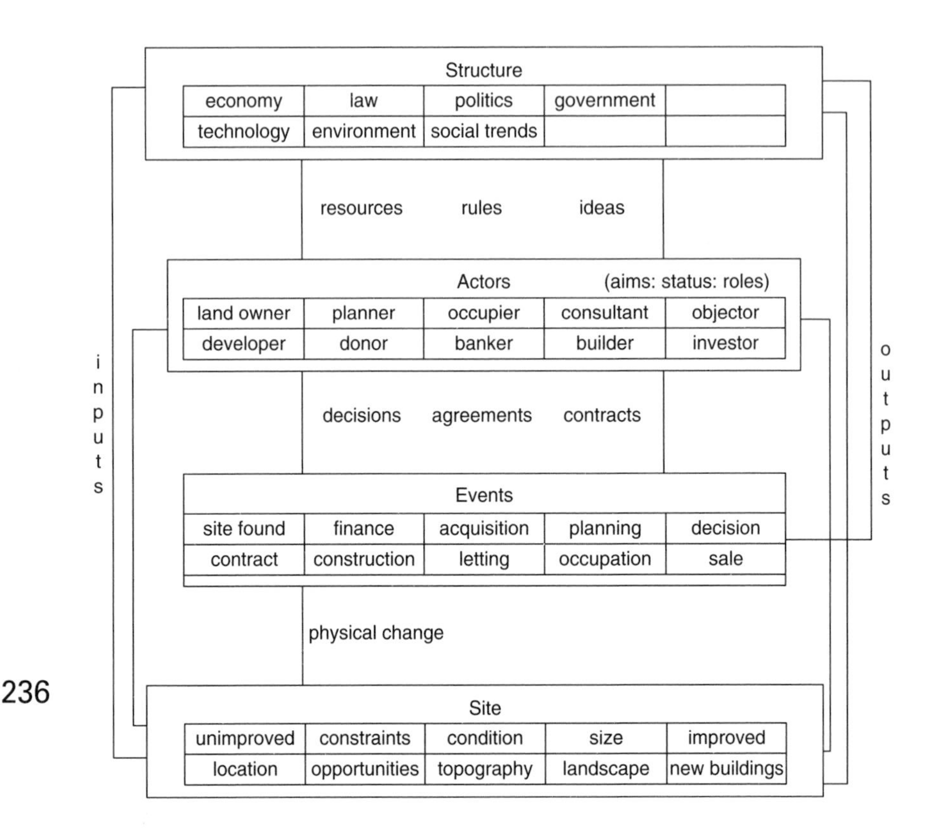

Figure 19.2 The Fisher and Collins model.
Source: based on Fisher and Collins (1999).

development and which inform and shape each other. Different interests involved may present or hold conflicting sets of views and priorities. Their judgements about impact or about viability may differ. What this means is that each development is likely to be moulded differentially depending on the resources and knowledges involved or which are 'circulating'. The outcome (i.e. the 'development outcome') represents an assemblage, or the result of a contestation process, whereby various forces have shaped actions.

In this view the development outcome represents the result of a filtered and socially constructed process. If certain conditions or forces are not conducive for key actors, such as the developer or the planning authority, the development will probably not occur, or the development outcome will satisfy another interest, and perhaps not delivering physical development or that particular manifestation of development. Such outcomes are a product of actor interests and behaviours which are recursively linked to the wider forces of which they are a part. In practice the different forces shown in Figure 19.3 overlap and interact (as shown by the 'recursively interrelating' arrows). They act to reinforce, contest and restructure the assumptions and implications of other forces. This implies a constantly dynamic set of development conditions which are reflected in terms of various kinds of development regulation, particular built forms or perhaps land use designations that act to shape development processes and outcomes. An example of this is the contrasting built forms and development outcomes produced as a result of corporate institutional and 'independent' capitalist investment actor networks and their related conditions (see Doak and Karadimitriou, 2007; Guy et al., 2002).

Similarly, if we consider 'sustainable development' in this light we might argue that particular alignments of forces have come about to deliver on particular actors' agendas. This has been achieved due to a complex assembly of actors and resources and possibly a series of

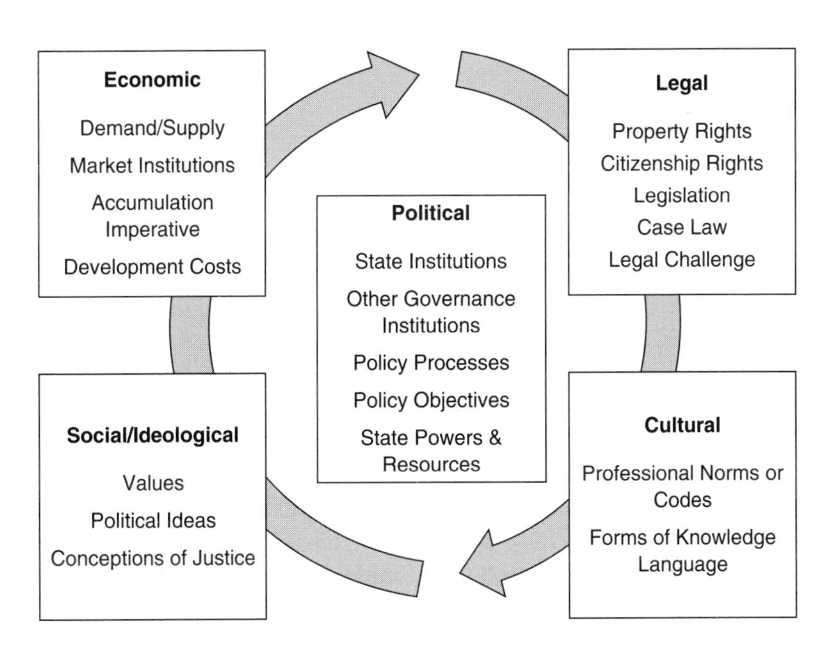

Figure 19.3 Forces acting to structure development processes.

'forced' compromises. In this way any development outcome may be viewed as a unique event, or unique manifestation of those relations.

The role of planning has been said to provide conditions whereby the range of actors can have some basis of certainty over each other's preferences and priorities. These should be represented through plans and policies in spatial terms, and the determination of planning applications provides a negotiative space to integrate different forces to some degree.

Scales of development

The fourth aspect is the various scales and layers through which development is orchestrated and implemented. This element is reasonably straightforward in comparison with the other elements already discussed. The existence of different 'layers' of development that take place when land or buildings are built or changed, range from the individual site to the national scale (and possibly beyond). This is both in terms of actual size or built coverage to the impacts that development has or the scale at which the development will provide benefits. Thus, the development of an individual site or building may have impacts upon, and implications for, the surrounding area, for the settlements in which it located and, through wider network relations, possibly for other places around the globe.

In this way a 'Russian doll' type picture may be visualized, in which development operations and impacts take effect at different spatial levels with one level recursively linked to the next. Thus, resources and materials are allocated and orchestrated from a range of places in order to build out a particular development scheme, but then that building and the uses/users within it have environmental, economic and social impacts, or externality effects, that ripple out across space and time. This is one of the complex considerations that planners are charged with, and which sets apart planners from developers, or other single interest actors involved with development. Planners are attempting to anticipate such outcomes and understand what is appropriate and in what location it may be best located, or where the negative externalities (Chapter 16) may be least harmful. This is also where the hierarchy of policy and decision-making influences different types and intensities of development due to the relative impact or significance of the development.

Forms of development

The fifth and last aspect or element that we discuss here is the multiple *forms* of development. These are driven by different needs and prioritisations shaped by the actors involved and through the combination of the other elements mentioned above. The *forms* are essentially products of the foregoing elements and reflect how these have come together to produce a particular outcome. The idea of a form of development is the particular manifestation that the development takes, which may be a large office block, a small housing estate, a new factory complex or indeed a mix of uses on site that are deemed to be in some way desirable. The outcome may not necessarily be a physical built form, however, it could be to maintain or create an open space for example. So in some cases physical development may not take place. Equally the less tangible development of community may have produced some effect but not one that can be easily seen or measured – the 'development' could therefore take the form of a change in attitude or a new set of relations being brokered. In terms of economic development this is best expressed as an aim rather than an outcome. Local economic development is a form of activity that centres on developing employment and economic activity and may involve both community and physical development. This may involve new built developments or other physical improvements, and may also comprise other measures such as place marketing or the use of fiscal incentives to promote 'economic development' of an area.

At its broadest, the forms of development reflect the outcomes of a contested process whereby decisions about land use and local environment are taken. This element in our analysis, when centring largely on physical development, refers to the different land uses and types of property as well as the functionality of the development. Different forms or types of development will be deemed more or less appropriate in different places and at different times. This may be due to some intrinsic quality of the area involved, or as with *scales* of development outlined above, it may be to do with the specific or likely externalities that different types of development produce (e.g. an industrial plant producing fumes, or a large office development producing many thousands of car trips per day), or to the presence of particular voices in the process or the economic conditions that prevail and structure development possibilities.

Given the variety of land uses which planning systems attempt to control or regulate, the categorisation of those uses can become a significant factor bearing on the location and form of development produced. In the UK context, a key tool to assist in this is the Use Classes Order (HMSO, 2010) which defines a set of land uses and sets up a need to obtain permission to change use, or indicates where a change of land use may not require permission. This approach makes use of a set of clearly defined land use classifications that operate to define and separate different uses which are seen to have varying implications and impacts when juxtapositioned with other uses and users. In many other countries this regulation of land use is achieved through zoning ordinances, in which the mix of uses and other development characteristics are specified in a detailed land use zoning plan of some kind.

Equally some variations in terms of the quantity and location of physical development may be related to a perceived need that has become pressing. In this situation the interface between the market acting to indicate demand and the state recognising or endorsing that demand (or need) may align and lead to big pushes where, for example, brownfield development is prioritised and housing in urban areas is encouraged by planning policy signals and with other incentives being provided. Sustainable development aims in such circumstances tend to be seen where new physical development is accompanied by public consultation and provision of community facilities (and possibly with greener development being brokered). However, the viability in market terms of such developments is susceptible to economic boom and bust cycles as experienced in 2008 in the UK and elsewhere around the world. This, once again, illustrates the fact that development involves the interaction of actors (in relation to other actors) with wider contextual forces that both constrain and facilitate their objectives.

Conclusion

The chapter has discussed a broad view of the concept of development which includes property development, economic development and community development, while linking this to overarching goals such as sustainable development as examined in Chapter 3. We can see how development impacts on quality of life and is orchestrated and affected by numerous factors or elements. Planning policy and practitioners may seek to play a coordinating role but formal planning is only one element in shaping or managing urban change. Our account of development as a process indicates how outcomes are complex and contingent on a set of conditions and relations regardless of whether we are focusing purely on 'property development' or on wider conceptualisations.

Such efforts to plan also reflect compromises, where competing forces and powers come together. Given the broader view of development that has been explained, we can perceive that questions of not only 'how and why' development occurs are important but 'when, what and where' it occurs. We can conclude that development is multifaceted in terms of actors, relations, forces, scales and forms; and this means that the role that planning and planners play in orchestrating development is as just one actor seeking to knit or weave other actors and forces together to provide acceptable – hopefully sustainable – physical and other development outcomes. This role as intermediary is an important and yet often poorly understood one.

FURTHER READING

Wilkinson and Reed (2008) provide a relatively simple introduction to the property development process, while Adams (1994) and Rydin (2003) contain more critically informed overviews. A number of writers have produced summaries and critiques of the different theoretical models of the property development process, including Ratcliffe (1978) and Fisher and Collins (1999). The first, more considered or broadened, attempt at this was Healey and Barratt's article in *Urban Studies* (1990) and this has been followed up by a procession of papers that have kept the debate going, adding new ideas and insights along the way (e.g. Ball, 1998, 2002; Doak and Karadimitriou, 2007; Fisher, 2005; Gore and Nicholson, 1991; Guy and Henneberry, 2000, 2002; Healey, 1992, 1992).

In widening the discussion of 'development' Midgley (1995) and Amin and Roberts (2008) both examine community development processes and issues, while the DTA (2009) provides a practice-based perspective on this. A comprehensive take on planning and community development is set out in Roseland (2000), with Woolcock (1998) discussing economic development and social capital together. The Blakely and Leigh (2010) textbook is one of the standard volumes available on local economic development and usefully provides theory and practice examples on this topic area.

REFERENCES

ACRE (2007) *Community and Neighbourhood Planning Toolkit*. Cirencester: Action with Communities in Rural England.

Adams, D. (1994) *Urban Planning and the Development Process*. London: UCL Press.

Adams, D. (1995) *The HitchHiker's Guide to the Galaxy: A Trilogy in Five Parts*. London: Heinemann.

Adams, D. and Watkins, C. (2002) *Greenfields, Brownfields and Housing Development*. London: John Wiley & Sons/Blackwell.

Adams, D., Watkins, C. and White, M. (eds) (2005) *Planning, Public Policy and Property Markets*. Oxford: Blackwell.

Agnew, J. (1987) *Place and Politics: The Geographical Mediation of State and Society*. London: Allen and Unwin.

Alasuutari, P. (1995) *Researching Culture: Qualitative Method and Cultural Studies*. London: Sage

Alexander, E.R. (2001) A transaction-cost theory of land use planning and development control: toward the institutional analysis of public planning, *Town Planning Review*, 72(1): 45–75.

Alexander, E.R. (2002a) The public interest in planning: from legitimation to substantive plan evaluation, *Planning Theory*, 1(3): 226–49.

Alexander, E.R. (2002b) Planning rights: towards normative criteria for evaluating plans, *International Planning Studies*, 7(3): 191–212.

Allen, T. (2005) *Property and the Human Rights Act 1998*. Oxford: Hart.

Allison, L. (1975) *Environmental Planning: A Political and Philosophical Analysis*. London: Allen and Unwin.

Allmendinger, P. (1997) *Thatcherism and Planning: The Case of Simplified Planning Zones*. Aldershot: Ashgate.

Allmendinger, P. (2001) *Planning in Postmodern Times*. London: Routledge.

Allmendinger, P. (2002) *Planning Theory*. Basingstoke: Palgrave Macmillan.

Allmendinger, P. (2007) Mobile phone mast development and the rise of third party governance in planning, *Planning Practice and Research*, 22(2): 177–96.

Allmendinger, P. (2009) *Planning Theory*, 2nd edn. Basingstoke: Palgrave Macmillan.

Alterman, R. (2001) *National Level Planning in Democratic Countries*. Liverpool: Liverpool University Press.

Amati, M. (ed.) (2008) *Urban Green Belts in the 21st Century*. London: Ashgate.

Amati, M. and Parker, G. (2007) Containing Tokyo's growth: the effect of land reform on the Green Belt in Japan, 1943–1970, in Miller C. and Roche, M. (eds), *Past*

Matters: Planning History and Heritage in the Pacific Rim. London: Cambridge Scholars Press.

Ambrose, P. (1986) *Whatever Happened to Planning?* London: Methuen.

Amin, A. and Roberts, J. (2008) Knowing in action: beyond communities of practice, *Research Policy*, 37(2): 353–69.

Amin, A. and Thrift, N. (1995) Globalisation, institutional 'thickness and the local economy', in Healey, P., Cameron, S., Davoudi, S., Graham, S. and Madani-Pour, A. (eds), *Managing Cities: The New Urban Context*. Chichester: John Wiley & Sons, pp. 91–108.

Anderson, B. (1991) *Imagined Communities: Reflections on the Origin and Spread of Nationalism*, 2nd edn. London: Verso.

Andrews, C. (2001) Analyzing quality-of-place, *Environment and Planning B: Planning and Design*, 28(2): 201–17.

Anheier, H., Gerhards, J. and Romo, F. (1995) Forms of capital and social structure in cultural fields: examining Bourdieu's social topography, *American Journal of Sociology*, 100(4): 859–903.

Appadurai, A. (1990) Disjuncture and difference in the global cultural economy, in Featherstone, M. (ed.), *Global Culture*. London: Sage.

Appadurai, A. (1996) *Modernity at Large: The Cultural Dimensions of Globalization*. Minneapolis: University of Minnesota Press.

Arefi, M. (1999) Non-place and placelessness as narratives of loss: rethinking the notion of place, *Journal of Urban Design*, 4(2): 179–93.

Armitage, D., Berkes, F. and Doubleday, N. (eds) (2007) *Adaptive Co-management*. Vancouver: University of British Columbia Press.

Arrowsmith, J., Sisson, K. and Marginson, P. (2004) What can benchmarking offer the open method of coordination? *Journal of European Public Policy*, 11: 311–28.

Ashworth, G. (1994) From history to heritage, from heritage to identity, in Ashworth, G. and Larkham, P. (eds), *Building a New Heritage. Tourism Culture and Identity in the New Europe*. London: Routledge, pp. 13–26.

Ashworth, G. (1998) Heritage, identity and interpreting a European sense of place, in Uzzell, D. and Ballantyne, R. (eds), *Contemporary Issues in Heritage and Environmental Interpretation: Problems and Prospects*. London: The Stationery Office, pp. 112–32.

Auge, M. (1995) *Non-Places: Introduction to an Anthropology of Supermodernity*. London: Verso.

Azuela, A. and Herrera, C. (2007) Taking land around the world: international trends in expropriation for urban and infrastructure projects, in Lall, S.V., Freire, M., Yuen, B., Rajack, R. and Helluin, J. (eds.) *Urban Land Markets: Improving Land Management for Successful Urbanization*. Dordrecht: Springer.

BAA (2008) *BAA Supports Third Runway for Heathrow*. Press release. London: British Airports Authority. Online: http://www.heathrowairport.com/about-us/media-centre/press-releases (accessed 5 May 2010).

Bachtler, J. and Turok, I. (eds) (1997) *The Coherence of EU Regional Policy: Contrasting Perspectives on the Structural Funds*. London: Jessica Kingsley.

Bailey, J. (1975) *Social Theory for Planning*. London: Routledge and Kegan Paul.

Ball, M. (1998) Institutions in British property research: a review, *Urban Studies*, 35(9): 1501–1517.

Ball, M. (2002) Cultural explanations of regional property markets: a critique, *Urban Studies*, 39(8): 1453–1469.

Banister, D. (2002) *Transport Planning*, 2nd edn. London: Routledge/Taylor and Francis.

Banister, D., Stead, D., Steen, P., Akerman, J., Dreborg, K., Nijkamp, P. and Schleicher-Tappeser, R. (2000) *European Transport Policy and Sustainable Mobility*. London: Spon.

Barclay, C. (2010) *Financing Infrastructure: The Community Infrastructure Levy*. House of Commons Library Paper, 10 May 2010. Online: http://www.parliament.uk/briefingpapers/commons/lib/research/briefings/snsc-03890.pdf (accessed 25 February 2011).

Barratt, S. and Fudge, C. (eds) (1981) *Policy and Action*. London: Methuen.

Barrow, C. (1997) *Environmental and Social Impact Assessment: An Introduction*. London: Arnold.

Bartik, T. and Smith, V. (1987) Urban amenities and public policy, in Mills, E.S. (ed.) *The Handbook of Regional and Urban Economics*, vol. 2. London: Elsevier.

Battram, A. (1998) *Navigating Complexity*. London: The Industrial Society.

Bauman, Z. (1998a) *Globalization: The Human Consequences*. Oxford: Polity Press.

Bauman, Z. (1998b) *Postmodernity*. Cambridge: Polity Press.

Bauman, Z. (2001) *Community*. Cambridge: Polity Press.

Beck, F.D. (2001) Do state-designated enterprise zones promote economic growth? *Sociological Inquiry*, 71: 508–32.

Becker, G.S. (1962) Investment in human capital: a theoretical analysis, *The Journal of Political Economy*, 70: 9–49.

Becker, L. C. (1977) *Property Rights: Philosophic Foundations*. London: Routledge and Kegan Paul.

Beddoe, M. and Chamberlin, A. (2003) Avoiding confrontation: securing planning permission for on-shore wind farm development in England, *Planning Practice and Research*, 18(1): 3–17.

Begg, I. (1999) Cities and competitiveness, *Urban Studies*, 36(5–6): 795–809.

Begg, I. (ed.) (2002) *Urban Competitiveness: Policies for Dynamic Cities*. Bristol: Policy Press.

Benditt, T. (1973) The public interest, *Philosophy and Public Affairs*, 2(3): 291–311.

Bentley, I., McGlynn, S. and Smith, G. (1985) *Responsive Environments: A Manual for Designers*. Oxford: Elsevier.

Berger, J. and Luckmann, M. (1966) *The Social Construction of Reality*. New York, NY: Doubleday.

Bickerstaff, K. and Walker, G. (2005) Shared visions, unholy alliances: power, governance and deliberative processes in local transport planning, *Urban Studies*, 42(12): 2123–44.

Bijker, W. and Law, J. (eds) (1992) *Shaping Technology-Building Society*. Cambridge, MA: MIT Press.

Blakely, E. and Leigh, N. (2010) *Planning Local Economic Development: Theory and Practice*, 4th edn. Washington, DC: Sage.

Blomley, N. (1994) *Law, Space and the Geographies of Power*. New York: Guilford Press.

Blunden, J. and Curry, N. (eds) (1989) *A Future for our Countryside*. Oxford: Basil Blackwell.

Bolan, R. (1983) The structure of ethical choice in planning practice, *Journal of Planning Education and Research*, 3(1): 23–34.

Bond, S. and Fawcett-Thompson, M. (2007) Public participation and new urbanism: a conflicting agenda? *Planning Theory and Practice*, 8(4): 449–72.

Bonnes, M. and Bonaiuto, M. (2002) Environmental psychology: from spatial-physical environment to sustianable development, in Bechtel, R. and Churchman, A. (eds), *Handbook of Environmental Psychology*. London: John Wiley & Sons, pp. 28–54.

Bonnes, M. and Secchiaroli, G. (1995) *Environmental Psychology. A Psycho-Social Introduction*. London: Sage.

Booher, D. and Innes, J. (2002) Network power in collaborative planning, *Journal of Planning Education and Research*, 21(3): 221–36.

Booth, P. (1996) *Controlling Development*. London: UCL Press.

Booth, P. (2002) From property rights to public control: The quest for public interest in the control of urban development, *Town Planning Review*, 73(2): 153–70.

Bourdieu, P. (1984) *Distinction: A Social Critique of the Judgement of Taste*. London: Routledge and Kegal Paul.

Bourdieu, P. (1986) The forms of capital, in Richardson, J.G. (ed.) *Handbook of Theory and Research for the Sociology of Education*. New York: Greenwood Press.

Bourdieu, P. (1989) *Distinction. A Social Critique of the Judgement of Taste*, 2nd edn. London: Routledge.

Bourdieu, P. (1994) Structures, habitus and practices, in Mommsen, W.J. (ed.) *The Polity Reader in Social Theory*. Cambridge: Polity Press, pp. 95–110.

Bourdieu, P. (2002) Habitus, in Hillier, J. and Rooksby, E. (eds), *Habitus: A Sense Of Place*. Aldershot: Ashgate, pp. 27–33.

Bourdieu, P. and Passeron, J.C. (1977) *Reproduction in Education, Society and Culture*. London: Sage.

Bowen, M., Salling, M., Haynes, K. and Cyran, E. (1995) Toward environmental justice: spatial equity in Ohio and Cleveland, *Annals of the Association of American Geographers*, 85(4): 641–63.

Breheny, M. (ed.) (1992) *Sustainable Development and Urban Form*. Oxford: Pion.

Breheny, M. and Hooper, A. (eds) (1985) *Rationality in Planning: Critical Essays on the Role of Rationality in Urban and Regional Planning*. Oxford: Pion.

Brenner, N. (1998) Global cities, glocal states: global city formation and state territorial restructuring in contemporary Europe, *Review of International Political Economy*, 5(1): 1–37.

Brenner, N. (2004) *New State Spaces: Urban Governance and the Rescaling of Statehood*. Oxford: Oxford University Press.

Bridge, G. (1997) Mapping the terrain of time-space compression: power networks in everyday life, *Environment and Planning D: Society and Space*, 15: 611–26.

Brindley, T., Rydin, Y. and Stoker, G. (1989) *Remaking Planning: Politics of Urban Change in the Thatcher Years*. London: Unwin Hyman.

Bromley, D. (1991) *Environment and Economy: Property Rights and Public Policy*. Oxford: Blackwell.

Brownill, S. and Carpenter, J. (2007) Increasing participation in planning: emergent experiences of the reformed planning system in England, *Planning Practice and Research*, 22(4): 619–34.

Brownill, S. and Parker, G. (2010) Why bother with good works? The relevance of public participation(s) in planning in a post-collaborative era, *Planning Practice and Research*, 25(3): 275–282.

Bruton, M. (ed.) (1984) *The Spirit and Purpose of Planning*. London: Hutchinson.

Bryman, A. (1988) *Quantity and Quality in Social Research*. London: Routledge.

Bulkeley, H. (2005) Reconfiguring environmental governance: towards a politics of scales and networks. *Political Geography*, 24(8): 875–902.

Budd, L. and Hirmis, A. (2004) Conceptual framework for regional competitiveness, *Regional Studies*, 38(9): 1015–28.

Bunce, M. (1994) *The Countryside Ideal*. London: Routledge.

Burt, R. (1992) *Structural Holes*. Cambridge, MA: Harvard University Press.

Burt, R. (2004) Structural holes and good ideas, *American Journal of Sociology*, 110(2): 349–99.

Byrne, D. (1998) *Complexity Theory and the Social Sciences*. London: Routledge.

Byrne, D. (2003) Complexity theory and planning theory: a necessary encounter, *Planning Theory* 2(3): 171–8.

CABE (2000) *By Design – Better Places to Live: A Companion Guide to PPG3*. London: Department of Transport, Local Government and the Regions.

Callon, M. (1991) Techno-economic networks and irreversibility, in Law, J. (ed.), *A Sociology of Monsters*. London: Routledge.

Callon, M. (1998) An essay on framing and overflowing: economic externalities revisited by sociology, in Callon, M. (ed.), *The Laws of the Markets*. Oxford: Blackwell, pp. 244–69.

Cameron, D. (2010) *Big Society Speech*, Liverpool, 19 July 2010. Online: http://www.number10.gov.uk/news/speeches-and-transcripts/2010/07/big-society-speech-53572 (accessed 15 December 2011).

Campaign to Protect Rural England (2011) Glossary of terms used on this site. Online: http://www.planninghelp.org.uk/resources/glossary (accessed 15 December 2011).

Campbell, H. (2006) Is the issue of climate change too big for spatial planning? *Planning Theory and Practice*, 7(2): 201–230.

Campbell, H. and Marshall, R. (2000) Moral obligations, planning, and the public interest; a commentary on current British practice, *Environment and Planning B: Planning and Design*, 27(2): 297–312.

Campbell, H. and Marshall, R. (2002) Utilitarianism's bad breath? A re-evaluation of the public interest justification in planning, *Planning Theory*, 1(2): 163–87.

Campbell, M. and Floyd, D. (1996) Thinking critically about environmental mediation, *Journal of Planning Literature*, 10(3): 235–47.

Campbell, S. and Fainstein, S. (eds) (2003) *Readings in Planning Theory*, 2nd edn. Malden, MA: Blackwell.

Cannon, J.Z. (2005) Adaptive management in superfund: learning to think like a contaminated site, *New York University Environmental Law Journal*, 13(3): 561–612.

Carmona, M. (2009) Sustainable urban design: principles to practice, *International Journal of Sustainable Development*, 12(1): 48–77.

Carmona, M. and Sieh, L. (2004) *Measuring Quality in Planning*. London: Taylor and Francis.

Carmona, M. and Sieh, L. (2005) Performance measurement innovation in English planning authorities, *Planning Theory and Practice*, 6(3): 303–33.

Carmona, M., Heath, T., Oc, T. and Tiesdall, S. (2003) *Public Places, Urban Spaces: The Dimensions of Urban Design*. London: Architectural Press.

Capra, F. (1982) *The Turning Point: Science, Society and the Rising Culture*. New York: Bantam.

Capra, F. (2002) *The Hidden Connections: A Science for Sustainable Living*. London: HarperCollins.

Castells, M. (1996) *The Rise of the Network Society*. Oxford: Blackwell.

Castree, N. (2003) Place: connections and boundaries in an inter-dependent world, in Holloway, S., Rice, S. and Valentine, G. (eds), *Key Concepts in Geography*. London: Sage, pp. 165–86.

Chadwick, G. (1978) *A Systems View of Planning*, 2nd edn. Oxford: Pergamon Press.

Cherry, G. (1974) *The Evolution of British Town Planning*. London: Leonard Hill.

Chettiparamb, A. (2007) Dealing with complexity: an autopoietic view of the People's Planning Campaign, Kerala, *Planning Theory and Practice*, 8(4): 489–508.

Cilliers, P. (1998) *Complexity and Postmodernism: Understanding Complex Systems*, London: Routledge.

Claydon, J. (1996) Negotiations in planning, in Greed, C. (ed.), *Implementing Town Planning*. Harlow: Longman, pp. 110–20.

Cloke, P. (ed.) (1987) *Rural Planning: Policy into Action?* London: Harper and Row.

Cochrane, A. (1986) Community politics and democracy, in Held, D. and Pollitt, C. (eds), *New Forms of Democracy*. London: Sage.

Cochrane, A. (2003) The new urban policy. Towards empowerment or incorporation? in Raco, M. and Imrie, R. (eds), *Urban Renaissance? New Labour Community and Urban Policy*. Bristol: Policy Press, pp. 223–34.

Cochrane, A. (2007) *Understanding Urban Policy: A Critical Approach*. Oxford: Blackwell.

Cockburn, C. (1977) *The Local State: Management of Cities and People*. London: Pluto Press.

Coleman, J. (1988) Social capital in the creation of human capital, *American Journal of Sociology*, 94: 95–120.

Coleman, J. (1990) *Foundations of Social Theory*. Cambridge, MA: Harvard University Press.

Coleman, R. (2002) *Revise PPG15! The Case for Changes to PPG 15*. London: Richard Coleman Consultancy.

Condon, P. (2007) *Design Charrettes for Sustainable Communities*. Washington, DC: Island Press.

Cooper, D. (1998) *Governing Out of Order: Space, Law and the Politics of Belonging*. London: Rivers Oram Press.

Cooper, L. and Sheate, W. (2002) Cumulative effects assessment: a review of UK environmental impact statements, *Environmental Impact Assessment Review*, 22(4): 415–39.

Corry, D. and Stoker, G. (2003) *New Localism Refashioning the Centre-local Relationship*. London: New Local Government Network.

Council of Europe. (2003) *European Conference of Ministers Responsible for Regional/Spatial Planning (CEMAT) – Overview Document*. Online: http://www.coe.int/t/dg4/cultureheritage/Source/Policies/CEMAT/CEMAT_leaflet_EN.pdf (accessed 15 December 2011).

Counsell, D. and Haughton, G. (2007) Spatial planning for the city-region, *Town and Country Planning,* 76(8): 248–51.

Crang, M. (1998) *Cultural Geography*. London: Routledge.

Cranston, M. (1973) *What Are Human Rights?* New York: Basic Books.

Crenson, M.A. (1971) *The Unpolitics of Air Pollution: A Study of Non-Decision Making in the Cities.* Baltimore, MD: Johns Hopkins University Press.

Cresswell, T. (2004) *Place: A Short Introduction.* Oxford: Blackwell.

Crouch, D. and Parker, G. (2003) Digging up utopia? *Geoforum*, 34(3): 395–408.

Crow, G. and Allan, G. (1994) *Community Life*. Wallingford: Harvester Wheatsheaf.

Cullen, G. (1961) *Townscape*. London: Architectural Press.

Cullingworth, J. (ed.) (1999) *Fifty Years of Urban and Regional Policy in the UK.* London: Athlone Press.

Cullingworth, B. and Caves, R. (2008) *Planning in the USA: Policies, Issues and Processes*, 2nd edn. London: Routledge.

Cullingworth, J. and Nadin, V. (2006) *Town and Country Planning in the UK*, 14th edn. Basingstoke: Palgrave Macmillan.

Cumbers, A. and MacKinnon, D. (2004) Introduction: clusters in economic development, *Urban Studies*, 41: 959–969.

Cypher, M.L. and Forgey, F.A. (2003) Eminent domain: an evaluation based on criteria relating to equity, effectiveness and efficiency, *Urban Affairs Review*, 39(2): 254–268.

Daft, R. (2009) *Organization Theory and Design*, 10th edn. Mason, OH: South Western Cengage Press.

Darlow, A., Percy-Smith, J. and Wells, P. (2007) Community strategies: are they delivering joined-up governance? *Local Government Studies*, 33(1): 117–29.

Davidoff, P. (1965) Advocacy and pluralism in planning, *Journal of the American Institute of Planners*, 31: 331–338.

Davies, A. (2002) Power, politics and networks: shaping partnerships for sustainable communities, *Area*, 34(2): 190–203.

Davies, J. (1972) *The Evangelistic Bureaucrat.* London: Tavistock.

DCLG (2005) *Planning Policy Statement 1.* London: Department of Communities and Local Government.

DCLG. (2007a) *Supplement to PPS1: Climate Change.* London: Department of Communities and Local Government.

DCLG (2007b) *Preparing Community Strategies: Government Guidance to Local Authorities,* London: Department for Communities and Local Government.

DCLG (2007c) *Planning for a Sustainable Future.* London: Department for Communities and Local Government.

DCLG (2008) *PPS12: Local Development Frameworks (Revised, June 2008).* London: Department for Communities and Local Government.

DCLG (2010) *Planning Policy Statement 5: Planning and the Historic Environment.* London: Department for Communities and Local Government.

DCLG (2011) *National Planning Policy Framework* (Draft). London: Department for Communities and Local Government.

DCLG (2012) *National Planning Policy Framework.* London: Department for Communities and Local Government.

de Roo, G. and Porter, G. (2007) *Fuzzy Planning: The Role of Actors in a Fuzzy Governance Environment*. Aldershot: Ashgate.

de Roo, G. and Silva, E. (ed.) (2010) *A Planners Meeting with Complexity*. Aldershot: Ashgate.

Dean, M. (1996) Foucault and the enfolding of government, in Barry, A., Osborne, T. and Rose, N. (eds), *Foucault and Political Reason*. London: UCL Press.

Deas, I. and Giordano, B. (2001) Conceptualising and measuring urban competitiveness in major English cities: an exploratory approach, *Environment and Planning A*, 33(8): 1411–29.

Delafons, J. (1997) *Politics and Preservation*. London: Routledge.

Delanty, G. (2003) *Community*. London: Routledge.

Demsetz, H. (1967) Towards a theory of property rights, *The American Economic Review*, 57(2): 347–59.

Denman, D.R. (1978) *Place of Property: New Recognition of the Function and Form of Property Rights in Land*. Berkhamsted: Geographical Publications.

Derwentside District Council (2008) *Refusal Of Planning: Permission Application Number 1/2008/0077/DMFP*. Consett: Derwentside District Council. Online: www.planning.derwentside.gov.uk/planning/08-0077/decision.pdf (accessed 15 December 2011).

DETR (2000) *Planning Policy Statement 3: Housing*. London: Department of the Environment, Transport and the Regions.

DfT (2009) *Guidance on Local Transport Plans*, July 2009. London: Department for Transport.

Dietz, T., Ostrom, E. and Stern, P. (2003) The struggle to govern the commons, *Science*, 302(5652): 1907–12.

Doak, A. and Karadimitriou, N. (2007) (Re)development, complexity and networks: A framework for research, *Urban Studies*, 44(2): 209–29.

Doak, A. and Parker, G. (2005) Networked space? The challenge of meaningful participation and the new spatial planning in England, *Planning Practice and Research*, 20(1): 23–40.

Dobson, A. (2007) *Green Political Thought*, 4th edn. London: Routledge

Dodge, M. and Kitchin, R. (2001) *Mapping Cyberspace*. London: Routledge.

DoE (1992) *Planning Policy Guidance Note 19: Outdoor Advertisement Control*. London: Department of the Environment.

DTA (2009) *Trust in Communities: A Manifesto from the Development Trusts Association*. London: Development Trusts Association.

DTI (1998) *Regional Competitiveness Indicators*. London: HMSO.

DTI (2004) *Regional Competitiveness and the State of the Regions*. London: HMSO.

Dunning, J. (2001) *Global Capitalism at Bay*? London: Routledge.

Dunning, J., Bannerman, E. and Lundan, S. (1998) *Competitiveness and Industrial Policy in Northern Ireland*. Research Monograph No. 5. Belfast: NI Research Council.

Duxbury, R.M.C. (2009) *Planning Law and Procedure*, 14th edn. Oxford: Oxford University Press.

ECC (1973) *Essex Design Guide for Residential Areas*. Chelmsford: Essex County Council.

Egan, J. (2004) *The Egan Review: Skills for Sustainable Communities*. London: Office of the Deputy Prime Minister.

Ekins, P., Simon, S., Deutsch L., Folke, C., and De Groot, R. (2003) A framework for the practical application of the concepts of critical natural capital and strong sustainability, *Ecological Economics*, 44: 165–185.

Elkington, J. (1994) Towards the sustainable corporation: win-win-win business strategies for sustainable development, *California Management Review*, 36(2): 90–100.

Ellis, H. (2002) Planning and public empowerment: third party rights in development control, *Planning Theory and Practice*, 1(2): 203–17.

Ellis, H. (2004) Discourses of objection: towards an understanding of third-party rights in planning, *Environment and Planning A*, 36(9): 1549–70.

Elson, M. (1986) *Greenbelts: Conflict Mediation in the Urban Fringe*. London: Heinemann.

Elster, J. (1998) *Deliberative Democracy*. Cambridge: Cambridge University Press.

English Heritage (2001) *Power of Place: The Future of the Historic Environment*. London: English Heritage.

English Heritage (2005) *Guidance on the Management of Conservation Areas*. London: English Heritage.

English Heritage (2010) *Conservation Areas at Risk. Frequently Asked Questions*. Online: http://www.english-heritage.org.uk/publications/faq-conservation-areas/faq.pdf (accessed 26 August 2010).

Ennis, F. (1997) Infrastructure provision, the negotiating process and the planner's role, *Urban Studies*, 34(12): 1935–54.

Evans, A. (2004) *Economics and Land Use Planning*. Oxford: Blackwell.

Evans, R., Guy, S. and Marvin, S. (1999) Making a difference: sociology of scientific knowledge and urban energy policies, *Science, Technology and Human Values*, 24(1): 105–31.

Fainstein, S. (2001) Competitiveness, cohesion, and governance: their implications for social justice, *International Journal for Urban and Regional Research*, 25(4): 884–8.

Faludi, A. (1973) *Planning Theory*. Oxford: Pergamon.

Faludi, A. (ed.) (2002) *European Spatial Planning*. Washington, DC: Lincoln Institute.

Farthing, S. and Ashley, K. (2002) Negotiation and the delivery of affordable housing through the English planning system, *Planning Practice and Research*, 17(1): 45–58.

Ferguson, M. (1980) *The Aquarian Conspiracy*. Los Angeles: Jeremy P. Tarcher.

Fine, B. (2001) *Social Capital Versus Social Theory: Political Economy and Social Science at the Turn of the Millennium*. London: Routledge.

Fischer, C.S. (1982) *To Dwell Among Friends: Personal Networks in Town and City*. Chicago: University of Chicago Press.

Fisher, P. (2005) The commercial property development process: case studies from Grainger Town, *Property Management*, 23(3): 158–175.

Fisher, P. and Collins, A. (1999) The commercial property development process, *Property Management*, 17(3): 219–230.

Fisher, R., Urry, W. and Patton, B. (1991) *Getting to Yes: Negotiating an Agreement Without Giving In*. London: Random House.

Flyvbjerg, B. (1996) The dark side of planning: rationality and realrationalität, in Mandelbaum, S., Mazza, L. and Burchell, R. (eds), *Explorations in Planning Theory*. New Brunswick, NJ: Center for Urban Policy Research, pp. 383–94.

Flyvbjerg, B. (1998) *Rationality and Power: Democracy in Practice*. Chicago, IL: Chicago University Press.

Foresight. (2010) *The Future of Land Use Report*. London: Defra/Communities and Local Government.

Forester, J. (1982) Planning in the face of power, *Journal of the American Planning Association*, 64(1): 67–80.

Forester, J. (1987) Planning in the face of conflict: negotiation and mediation strategies in local land use regulation, *Journal of the American Planning Association*, 53(3): 303–14.

Forester, J. (1989) *Planning in the Face of Power*. Cambridge, MA: MIT Press.

Forester, J. (1993) *Critical Theory, Public Policy and Planning Practice*. Albany, NY: State University of New York Press.

Forester, J. (1999) *The Deliberate Practitioner. Encouraging Participatory Planning Processes*. Cambridge, MA: MIT Press.

Forrest, R. and Kearns, A. (2001) Social cohesion, social capital and the neighbourhood, *Urban Studies*, 38(12): 2125–2143.

Foucault, M. (1970) *The Order of Things*. London: Tavistock.

Fowler, A. (1990) *Negotiation Skills and Strategies*. London: IPM.

Freeden, M. (1991) *Rights*. Milton Keynes: Open University Press.

Friedmann, J. (1973) The public interest and community participation: towards a reconstruction of public philosophy, *Journal of the American Institute of Planners*, 39(1): 2–12.

Friedmann, J. (1993) Toward a non-euclidian mode of planning, *Journal of the American Planning Association*, 59(4): 482–5.

Friedmann, J. (1998) Planning theory revisited, *European Planning Studies*, 6(3): 245–53.

Friedrich, C. (ed.) (1962) *The Public Interest*. New York, NY: Atherton Press.

Gallent, N., Juntti, M., Kidd, S. and Shaw, D. (2008) *Introduction to Rural Planning*. Abingdon: Routledge.

Gans, H. (1969) Planning for people not buildings, *Environment and Planning A*, 1: 33–46.

Geddes, P. (1915) *Cities In Evolution: An Introduction To The Town Planning Movement And To The Study Of Civics*. London: Benn.

Geisler, C. and Daneker, G. (eds) (2000) *Property and Values*. Washington, DC: Island Press.

Gelfand, M.J. and Brett, J.M. (eds) (2004) *Handbook of Negotiation and Culture*. Palo Alto, CA: Stanford University Press.

Ghezzi, S. and Mingione, E. (2007) Embeddedness, path dependency and social institutions. An economic sociology approach, *Current Sociology*, 55(1): 11–23.

Ghimire, K. and Pimbert, M. (eds) (1997) *Social Change and Conservation*. London: Earthscan.

Giddens, A. (1984) *The Constitution of Society: Outline of the Theory of Structuration*. Cambridge: Polity Press.

Gilpin, A. (1995) *Environmental Impact Assessment. Cutting Edge for the Twenty-First Century*. Cambridge: Cambridge University Press.

GLA (2008) *Who Gains? The Operation of Section 106 Planning Agreements in London*. London: Greater London Assembly Online: http://legacy.london.gov.uk/assembly/reports/plansd/section-106-who-gains.pdf (accessed 15 December 2011).

Glasson, J. (1978) *An Introduction to Regional Planning*. London: Hutchinson.

Glasson, J. and Marshall, T. (2007) *Regional Planning*. London: Routledge.

Glasson, J., Therivel, R. and Chadwick, A. (2005) *Introduction to Environmental Impact Assessment*. London: Routledge.

Goldthorpe, J., Llewellyn, C. and Payne, C. (1987) *Social Mobility and Class Structure in Modern Britain*, 2nd edn. Oxford: Oxford University Press.

Goodchild, R. and Munton, R. (1985) *Development and the Landowner*. London: Allen and Unwin.

Goodey, B. (1998) Essex design guide revisited, *Town and Country Planning*, 67(5): 176–8.

Gore, T. and Nicholson, D. (1991) Models of the land-development process: a critical review, *Environment and Planning A*, 23(5): 705–30.

Graham, B. (2002) Heritage as knowledge: capital or culture? *Urban Studies*, 39: 1003–17.

Graham, S. and Healey, P. (1999) Relational concepts of space and place: issues for planning theory and practice, *European Planning Studies*, 7(5): 623–46.

Granovetter, M. (1985) Economic action and social structure: the problem of embeddedness, *American Journal of Sociology*, 91(3): 814–41.

Granovetter, M. (2005) The impact of social structure on economic outcomes, *The Journal of Economic Perspectives*, 19(1): 33–50.

Grant, M. (1999) Compensation and betterment, in Cullingworth, J. (ed.), *Fifty Years of Urban and Regional Policy in the UK*. London: Athlone Press, pp. 62–90.

Greed, C. (ed.) (1996) *Implementing Town Planning: The Role of Town Planning in the Development Process*. Longman: Harlow, Essex.

Greed, C. (1999) *Social Town Planning*. London: Routledge.

Guy, S. and Henneberry, J. (2000) Understanding urban development processes: integrating the economic and the social in property research, *Urban Studies*, 37(13): 2399–416.

Guy, S. and Henneberry. J. (2002) Bridging the Divide? Complementary Perspectives on Property, *Urban Studies*, 39(8): 1471–1478.

Guy, S., Henneberry, J., and Rowley, S. (2002) Development cultures and urban regeneration', *Urban Studies*, 39(7): 1181–1196.

Hague, C. and Jenkins, P. (eds) (2005) *Place Identity, Participation and Planning*. London: Routledge.

Hall, P. (2000) *Cities of Tomorrow*, 3rd edn. Oxford: Blackwell.

Hall, P. (2002) *Urban and Regional Planning*, 4th edn. London: Routledge.

Hall, P. and Pain, K. (2006) *The Polycentric Metropolis: Learning from Mega-City Regions in Europe*. London: Earthscan.

Hall, P. and Tewdwr-Jones, M. (2011) *Urban and Regional Planning*. 5th edn, Routledge: London.

Ham, C. and Hill, M. (1993) *The Policy Process in the Modern Capitalist State*, 2nd edn. Hemel Hempstead: Harvester Wheatsheaf.

Hambleton, R. (1986) *Rethinking Policy Planning*. Bristol: Policy Press.

Hamm, B. and Muttagi, P. (eds) (1998) *Sustainable Development and the Future of Cities*. London: Intermediate Technology Publications.

Hanley, N. and Barbier, E. (2009) *Pricing Nature: Cost-Benefit Analysis and Environmental Policy*. Cheltenham: Edward Elgar.

Hansen, S. (1991) *Comparing Enterprise Zones to Other Economic Development Techniques*. Newberry Park, CA: Sage.

Harvey, D. (1989) *The Condition of Postmodernity*. Oxford: Blackwell.

Hastings, J. and Thomas, H. (2005) Accessing the nation: disability, political inclusion and built form, *Urban Studies*, 42(3): 527–544.

Haughton, G. (1999) Environmental justice and the sustainable city, *Journal of Planning Education and Research*, 18(3): 233–43.

Haughton, G., Allmendinger, P., Counsell, D., Vigar, G. (2009) *The New Spatial Planning: Territorial Management with Soft Spaces and Fuzzy Boundaries*. London: Routledge.

Haus, M., Heinelt, H. and Stewart, M. (eds) (2005) *Urban Governance and Democracy*. London: Routledge.

Hawkes, J. (2001) *The Fourth Pillar of Sustainability: Culture's Essential Role in Public Planning*. Melbourne: Common Ground.

Healey, J. (2007) Nationally significant infrastructure – speech made by John Healey, on 30 October 2007, to the CBI Major infrastructure conference. Online: http://www.communities.gov.uk/speeches/corporate/cibinfrastructure (accessed 25 February 2011).

Healey, P. (1983) *Local Plans in British Land Use Planning*. Oxford: Pergamon Press.

Healey, P. (1990) Places, people and politics: plan making in the 1990s, *Local Government Policy Making*, 17(2): 29–39.

Healey, P. (1992) An institutional model of the development process, *Journal of Property Research*, 9: 33–44.

Healey, P. (1998) Building institutional capacity through collaborative approaches to urban planning, *Environment and Planning A*, 30(9):1531–1546.

Healey, P. (1999) Sites, jobs and portfolios: economic development discourses in the planning system, *Urban Studies*, 36(1): 27–42.

Healey, P. (2004) The treatment of space and place in the new strategic spatial planning in Europe, *International Journal of Urban and Regional Research*, 28(1): 45–67.

Healey, P. (2005) *Collaborative Planning. Shaping Place in a Fragmented World*, 2nd edn. Basingstoke: Palgrave Macmillan.

Healey, P. (2006) *Urban Complexity and Spatial Strategies: Towards a Relational Planning for Our Times*. London: RTPI.

Healey, P. and Barrett, S. (1990) Structure and agency in land and property development processes: some ideas for research, *Urban Studies*, 27(1): 89–104.

Healey, P., McNamara, P., Elson, M. and Doak, A. (1988) *Land Use Planning and the Mediation of Urban Change: The British Planning System in Practice*. Cambridge: Cambridge University Press.

Healey, P., Purdue, M. and Ennis, F. (1995) *Negotiating Development: Rationales and Practice for Development Obligations*. London: E and F.N. Spon.

Heelas, P., Lash, S. and Morris, P. (eds) (1996) *Detraditionalization: Critical Reflections on Authority and Identity*. Oxford: Blackwell.

Henry, N. and Pinch, S. (2001) Neo-Marshallian nodes, institutional thickness, and Britain's 'motor sport valley': thick or thin? *Environment and Planning A*, 33(7): 1169–83.

Hill, M. (2005) *The Public Policy Process*, 4th edn. London: Pearson.

Hillier, J. (1999) Habitat's habitus: nature as sense of place in land use planning decision-making, *Urban Policy and Research*, 17(3): 191–204.

Hillier, J. (2000) Going round the back? Complex networks and informal action in local planning processes, *Environment and Planning A*, 32(1): 33–54.

Hillier, J. (2002) Direct action and agonism in planning practice, in Allmendinger, P. and Tewdwr-Jones, M. (eds), *Planning Futures*. London: Routledge, pp. 110–35.

Hillier, J. (2007) *Stretching Beyond the Horizon. A Multiplanar Theory of Spatial Planning and Governance*. Aldershot: Ashgate.

Hillier, J. and Healey, P. (eds) (2008) *Critical Essays in Planning Theory* (3 Volumes). Aldershot: Ashgate.

Hillier, J. and Rooksby, E. (eds) (2002) *Habitus: A Sense of Place*. Aldershot: Ashgate.

Hillier J. and Rooksby, E. (eds) (2005) *Habitus: A Sense of Place*, 2nd edn. Aldershot: Ashgate.

HMSO (1990) *Town and Country Planning Act 1990*. London: HMSO.

HMSO (2004) *2004 Planning and Compensation Act*. London: HMSO.

HMSO (2010) *The Town and Country Planning (Use Classes) (Amendment) (England) Order 2010 (SI 2010/653)*. London: HMSO.

HM Treasury (2004) *Devolving Decision Making: Meeting the Regional Economic Challenge: Increasing Regional and Local Flexibility*. London: H.M. Treasury.

Hobsbawm, E. (1994) *The Age of Extremes*. New York, NY: Pantheon.

Hoch, C. (1996) A pragmatic inquiry about planning and power, in Seymour, J., Mandelbaum, L. and Burchell, R. (eds) *Explorations in Planning Theory*, New Brunswick, NJ: Center for Urban Policy Research, pp. 30–44.

Hodge, I. (1999) Countryside planning: from urban containment to sustainable development, in Cullingworth, B. (ed.) *British Planning: 50 Years of Urban and Regional Policy*, London: The Athlone Press, pp. 91–104.

Holloway, S., Rice, S. and Valentine, G. (eds) (2009) *Key Concepts in Geography*, 2nd edn. London: Sage.

Honoré, A.M. (1961) Ownership, in Guest, A.G. (ed.) *Oxford Essays in Jurisprudence*. Oxford: Clarendon Press, pp. 107–147.

Howard, E., Hall, P., Ward, C. and Hardy, D. (2003) *Tomorrow: A Peaceful Path to Real Reform*. London: Routledge.

Howard, P. (2004) Spatial planning for landscape: mapping the pitfalls, *Landscape Research*, 29(4): 423–34.

Howe, E. (1992) Professional roles and the public interest in planning, *Journal of Planning Literature*, 6(3): 230–48.

Howe, J. and Langdon, C. (2002) Towards a reflexive planning theory, *Planning Theory*, 1(3): 209–25.

Hubbard, P., Kitchin, R. and Valentine, G. (eds) (2004) *Key Thinkers on Space and Place*. London: Sage.

Huggins, R. (2003) Creating a UK competitiveness index: regional and local benchmarking, *Regional Studies*, 36(1): 89–96.

Huxley, M. and Yiftachel, O. (2000) New paradigm or old myopia? Unsettling the communicative turn in planning theory, *Journal of Planning Education and Research*, 14(3):163–166.

IDeA (2010) *What are City Regions?* Improvement and Development Agency. Online: http://www.idea.gov.uk/idk/core/page.do?pageId=7773100 (accessed 15 December 2011).

Imrie, R. and Raco, M. (eds) (2003) *Urban Renaissance? New Labour, Community and Urban Policy*. Bristol: Policy Press.

Innes, J.E. (1995) Planning theory's emerging paradigm: Communicative action and interactive practice, *Journal of Planning Education and Research*, 14(3): 183–189.

Innes, J.E. and Booher, D.E. (1999) Consensus building and complex adaptive systems: a framework for evaluating collaborative planning, *Journal of the American Planning Association*, 65(4): 412–23.

Innes, J.E. and Booher, D.E. (2003) Collaborative policymaking: Governance through dialogue, in Hajer, M.A. and Wagenaar, H. (eds), *Deliberative Policy Analysis: Understanding Governance in the Network Society*. London: Cambridge University, pp. 33–59.

Innes, J.E. and Booher, D.E. (2010) *Planning with Complexity: An Introduction to Collaborative Rationality for Public Policy*. London: Routledge.

Jacobs, J. (1961) *Death and Life of Great American Cities*. New York: Random House.

Jansson, M., Goosen, H. and Omtzigt, N. (2006) A simple mediation and negotiation and support tool for water management in the Netherlands, *Landscape and Urban Planning*, 78(1): 71–84.

Jenkins, R. (1992) *Pierre Bourdieu*. London: Routledge.

Jenks, M. (ed.) (2005) *Future Forms and Design for Sustainable Cities*. London: Architectural Press.

Jenks, M. and Burgess, R. (eds) (2000) *Compact Cities: Sustainable Urban Forms for Developing Countries*. London: Spon Press.

Jiven, G. and Larkham, P. (2003) Sense of place, authenticity and character: a commentary, *Journal of Urban Design*, 8(1): 67–81.

Johnson, R. (1993) *Negotiation Basics*. London: Sage.

Jordan, G. (1990) Sub-governments, policy communities and networks, *Journal of Theoretical Politics*, 2(3): 319–38.

Kambites, C. and Owen, S. (2006) Renewed prospects for green infrastructure planning in the UK, *Planning Practice and Research*, 21(4): 483–96.

Khakee, A. (2002) Assessing institutional capital building in a Local Agenda 21 process in Goteborg, *Planning Theory & Practice*, 3(1): 53–68.

Kipfer, S. and Keil, R. (2002) Toronto Inc? Planning the competitive city in the new Toronto, *Antipode*, 34(2): 227–64.

Kitchen, T. and Whitney, D. (2004) Achieving more effective public engagement within the English planning system, *Planning Practice and Research*, 19(4): 393–413.

Kitson, M., Martin, R. and Tyler, P. (2004) Regional competitiveness: and elusive yet key concept, *Regional Studies*, 38: 991–99.

Klosterman, R. (1985) Arguments for and against planning, *Town Planning Review*, 56(1): 5–20.

Koresawa, A. and Konvitz, J. (eds) (2001) *Towards a New Role for Spatial Planning*. Paris: OECD.

Knorr-Cetina, K. (1999) *Epistemic Cultures: How the Sciences Make Knowledge*. Cambridge, MA: Harvard University Press.

Kropotkin, P. (1912) *Fields, Factories and Workshops*. London: Thomas Nelson.

Krugman, P. (1996) Making sense of the competitiveness debate, *Oxford Review of Economic Policy*, 12: 17–35.

Kwa C. (2002) Romantic and Baroque conceptions of complex wholes in the sciences, in Law, J. and Mol, A. (eds) *Complexities: Social Studies of Knowledge Practices*. Durham, NC and London: Duke University Press, pp 23–52.

Lai, L.W.C. (1994) The economics of land use zoning: a literature review and analysis of the work of Coase, *Town Planning Review*, 65(1): 77–98.

Lai, L.W.C. (2002) Libertarians on the road to town planning, *Town Planning Review*, 73(3): 289–310.

Lake, R. (1996) Volunteers, NIMBYs and environmental justice: dilemmas of democratic practice, *Antipode*, 28(2): 160–74.

Lambert, C. (2006) Community strategies and spatial planning in England: The challenges of integration, *Planning Practice and Research*, 21(2): 245–55.

Larkham, P. (1996) *Conservation and the City*. London: Routledge.

Latham, A., McCormack, D., McNeill, D. and McNamara, K. (2009) *Key Concepts in Human Geography*. London: Sage.

Laurini, R. (2001) *Information Systems for Urban Planning: A Hypermedia Cooperative Approach*. London: Taylor and Francis.

Law, J. (1999) After ANT: complexity, naming and topology, in Hassard, J. and Law, J. (eds) *Actor-Network Theory and After*. Oxford: Blackwell Publishers.

Leeds City Region (2010) *Leeds City Region Webpages*. Online: http://www.leedscity-region.gov.uk/ (accessed 30 August 2010).

Lever, W. and Turok, I. (1999) Competitive cities: an introduction to the review, *Urban Studies*, 36(5–6): 791–3.

Lewin, L. (1991) *Self-Interest and Public Interest in Western Politics*. Oxford: Oxford University Press.

Lin, N. (2001) *Social Capital: A Theory of Social Structure and Action*. Cambridge: Cambridge University Press.

Lin, N. and Erikson, B. (eds) (2008) *Social Capital: An International Research Program*. Oxford: Oxford University Press.

Lindblom, C. (1959) The science of muddling through, *Public Administration Review*, 19(2): 79–88.

Lindblom, C. (1965) *The Intelligence of Democracy*. New York, NY: Free Press.

Litman, T. and Burwell, D. (2006) Issues in sustainable transportation, *International Journal of Global Environmental Issues*, 6(4): 331–47.

Lovelock, J. (1979) *Gaia: A New Look at Life on Earth*. Oxford: Oxford University Press.

Low, N. (1991) *Planning, Politics and the State*. London: Unwin Hyman.

Lowndes, V. and Skelcher, C. (1998) The dynamics of multi-organizational partnerships: an analysis of changing modes of governance, *Public Administration*, 76: 313–333.

Luhmann, N. (1986) The autopoiesis of social systems, in Geyer F. and van der Zouwen, J. (eds) *Sociocybernetic Paradoxes: Observation, Control and Evolution of Self Steering Systems*, London: Sage, pp.172–192.

Luhmann, N. (1995). *Social Systems*. Stanford, CA: Stanford University Press.

Lundblom, C. (1959) 'The science of muddling through', *Public Administration Review*, 19: 79–88.

Lundvall, B.Å. and Maskell, P. (2000) Nation states and economic development: from national systems of production to national systems of knowledge creation and learning, in Clark, G.L., Feldman, M.P and Gertler, M.S. (eds) *The Oxford Handbook of Economic Geography*. New York: Oxford University Press.

Lynch, K. (1960) *The Image of the City*. Cambridge, MA: MIT Press.

Lynch, K. (1981) *A Theory of Good City Form*. Cambridge, MA: MIT Press.

Mackay, H. (1996) *Swim With the Sharks Without Being Eaten Alive*. New York, NY: Ballantine Books.

Madanipour, A. (1996) *Design of Urban Space: An Inquiry into a Socio-Spatial Process*. Chichester: John Wiley & Sons.

Maddux, R. (1999) *Successful Negotiation*. London: Kogan Page.

Majone, G. and Wildavsky, A. (1978) Implementation as evolution, *Policy Studies Review Annual,* 2: 103–117.

Malecki, E. (2007) Cities and regions competing in the global economy: knowledge and local development policies, *Environment and Planning C*, 25: 638–54.

Marcuse, P. (1976) Professional ethics and beyond: values in planning, *Journal of the American Institute of Planning*, 42(3): 264–74.

Markusen, A. (1996) Sticky places in slippery space: a typology of industrial districts, *Economic Geography*, 72: 293–313.

Marsh, D. (1998) *Comparing Policy Networks*. Buckingham: Oxford University Press.

Marsh, D. and Rhodes, R. (1992) *Policy Communities and Issue Networks: Beyond Typology*. Oxford: Clarendon.

Marsh, D. and Smith, M. (2000) Understanding policy networks, *Political Studies*, 48(1): 4–21.

Marshall, T.H. and Bottomore, T.B. (1992) *Citizenship and Social Class*. London: Pluto Press.

Martin, R. and Sunley, P. (2003) Deconstructing clusters: chaotic concept or policy panacea? *Journal of Economic Geography*, 3: 5–35.

Massey, D. (2005) *For Space*. London: Sage.

Massey, D.B. and Catalano, A. (1978) *Capital and Land: Landownership by Capital in Great Britain*. London: Edward Arnold.

Maurici, J. (2002) Human rights update: part 1, *The Planning Inspectorate Journal,* 25: 12–16.

Maurici, J. (2003) Human rights update: part 2, *The Planning Inspectorate Journal,* 26: 9–15.

Mazmanian, D. and Sabatier, P. (1989) *Implementation and Public Policy*. Lanham, MD: University Press of America.

McDowell, L. (ed.) (1997) *Undoing Place? A Geographical Reader*. New York, NY: John Wiley & Sons.

McLaughlin, M. (1987) Learning from experience: Lessons from policy implementation, *Educational Evaluation and Policy Analysis*, 9(2): 171–78.

McLoughlin, J.B. (1969) *Urban and Regional Planning: A Systems Approach*. London: Faber and Faber.

Meadowcroft, J. (2000) Sustainable development: a new(ish) idea for a new century? *Political Studies*, 48: 370–87.

Meadows, D. (1991) Let's have a little more feedback, *The Donnela Meadows Archive: Voice of a Global Citizen*. Online: http://www.sustainer.org/dhm_archive/index.php?display_article=vn311feedbacked (accessed 30 August 2011).

Meadows, D., Meadows, D., Randers, J. and Behrens, W. (1972) *The Limits to Growth: A Report for the Club of Rome on the Predicament of Mankind*. New York, NY: Universe Books.

Meadows, D., Randers, J. and Meadows, D. (2004) *Limits to Growth: The 30 Year Update*. Vermont: Chelsea Green Publishers.

Midgley, J. (1995) *Social Development: The Developmental Perspective in Social Welfare*. London: Sage.

Mill, J.S. (1859) On Liberty, in Collini, S. (ed.) (1989) *On Liberty and Other Writings*. Cambridge: Cambridge University Press.

Mol, A. and Law, J. (eds) (2002) *Complexities: Social Studies of Knowledge Practices*. Durham, NC: Duke University Press.

Molotch, H., Freudenburg, W. and Paulsen, K.E. (2000) History repeats itself, but how? City character, urban tradition, and the accomplishment of place, *American Sociological Review*, 65(6): 791–823.

Montgomery, J. (2007) *The New Wealth of Cities: City Dynamics and the Fifth Wave*. Aldershot: Ashgate.

Moore, V. (2010) *A Practical Approach to Planning Law*, 11th edn. Oxford: Oxford University Press.

Morphet, J. (2010) *Effective Practice in Spatial Planning*. Abingdon: Routledge.

Morris, E. (1997) *British Town Planning and Urban Design: Principles and Policies*. Longman: Harlow.

Morris, P. and Therivel, R. (eds) (2001) *Methods of Environmental Impact Assessment*, 2nd edn. London: Spon.

Moulaert, F. and Cabaret, K. (2006) Planning, networks and power relations: is democratic planning under capitalism possible? *Planning Theory*, 5: 51–70.

Munro, R. and Mouritsen, J. (eds), *Accountability: Power Ethos and the Technologies of Managing*. London: Thomson Business Press, pp. 283–305.

Murdoch, J. (1997a) The shifting territory of government: some insights from the rural white paper, *Area*, 29(2): 109–18.

Murdoch, J. (1997b) Towards a geography of heterogeneous associations. *Progress in Human Geography*, 21(3): 321–37.

Murdoch, J. (1998) The spaces of actor-network theory, *Geoforum*, 29(4): 357–74.

Murdoch, J. (2004) Putting discourse in its place: planning, sustainability and the urban capacity study, *Area*, 36(1): 50–8.

Murdoch, J. (2005) *Post-structuralist Geography: A Guide to Relational Space*. London: Sage.

Murdoch, J. and Abram, S. (2002) *Rationalities of Planning: Development Versus Environment in Planning for Housing*. Basingstoke: Ashgate.

Mynors, C. (2006) *Listed Buildings, Conservation Areas and Monuments*, 4th edn. London: Sweet & Maxwell.

Nadin, V. (2007) The emergence of spatial planning in England, *Planning Practice and Research*, 22(1): 43–62.

Naess, A. (1989) *Ecology, Community and Lifestyle: Outline of an Ecosophy*. Cambridge: Cambridge University Press.

Nisbet, R. (1973) *The Social Philosophers: Community and Conflict in Western Thought*. New York, NY: Crowell.

Norberg-Schulz, C. (1980) *Genius Loci: Towards a Phenomenology of Architecture*. New York, NY: Rizzoli.

North, D. (1990) *Institutions, Institutional Change and Economic Performance*. Cambridge: Cambridge University Press.

Northern Way. (2010) *Moving Forward: The Northern Way*. Online: http://www.thenorthernway.co.uk/ (accessed 30 August 2010).

ODPM (2000a) *Our Towns and Cities: The Future – Delivering an Urban Renaissance*. (The Urban White Paper). London: The Stationery Office.

ODPM (2000b) *Environmental Impact Assessment: A Guide to Procedures*. London: Office of the Deputy Prime Minister.

ODPM (2001) *Planning Policy Guidance Note 2: Green Belts*. London: Office of the Deputy Prime Minister.

ODPM (2003a) *The Relationship Between Community Strategies and Local Development Frameworks*. London: Office of the Deputy Prime Minister.

ODPM (2003b) *Cities, Regions and Competitiveness*. London: Office of the Deputy Prime Minister.

ODPM (2004a) *Creating Local Development Frameworks: A Companion Guide to PPS 12*. London: Office of the Deputy Prime Minister.

ODPM (2004b) *Planning Policy Statement 12: Local Development Frameworks*. London: Office of the Deputy Prime Minister.

ODPM (2004c) *Competitive European Cities: Where do the Core Cities Stand?* Urban Research Paper No. 13. London: Office of the Deputy Prime Minister.

ODPM (2004d) *Planning Policy Statement 11: Regional Spatial Strategies*. London: Office of the Deputy Prime Minister.

ODPM (2004e) *Planning Policy Statement 7: Sustainable Development in Rural Areas*. London: Office of the Deputy Prime Minister.

ODPM (2005a) *Diversity and Planning. A Good Practice Guide*. London: Office of the Deputy Prime Minister.

ODPM (2005b) *Planning Policy Statement 1: Delivering Sustainable Development*. London: Office of the Deputy Prime Minister.

ODPM (2005c). *Circular 05/2005: Planning Obligations*, London: ODPM.

Oinas, P. and Malecki, E. (2002) The evolution of technologies in time and space: from national and regional to spatial innovation systems, *International Regional Science Review*, 25: 102–31.

Olsson, A. (2009) Relational rewards and communicative planning: understanding actor motivation, *Planning Theory*, 8(3): 263–81.

Ostrom, E. (1990) *Governing the Commons: the Evolution of Institutions for Collective Action*. Cambridge: Cambridge University Press.

Ostrom, E. (2003) How types of goods and property rights jointly affect collective action, *Journal of Theoretical Politics*, 15(3): 239–70.

Ouf, A. (2001) Authenticity and sense of place in urban design, *Journal of Urban Design*, 6(1): 73–86.

Owen, S. (1998) The role of village design statements in fostering a locally responsive approach to village planning and design in the UK, *Journal of Urban Design*, 3(3): 359–80.

Owen, S. (2002) From village design statements to Parish plans: pointers towards community decision-making in the planning system, *Planning Practice and Research,* 17(1): 3–16.

Owens, S. (2004) Siting, sustainable development and social priorities, *Journal of Risk Research* 7(2): 101–114.

Owens, S. and Cowell, R. (2002) *Land and Limits: Interpreting Sustainability in the Planning Process*. London: Routledge.

Paddison, R. (2001) *Communities in the City, Handbook of Urban Studies*. London: Sage.

Pahl, R. (1970) *Whose City?* Harmondsworth: Penguin.

Pandit, R. (2009) *Building World-Class Infrastructure For Competitiveness*, Financial Express, 7 January 2009. Online: http://www.financialexpress.com/news/building-worldclass-infrastructure-for-competitiveness/407595/ (accessed 15 December 2011).

Parker G. (2001) Planning and rights: some repercussions of the Human Rights Act 1998 for the UK, *Planning Practice and Research*, 16(1): 5–8.

Parker, G. (2002) *Citizenships, Contingency and the Countryside*. London: Routledge.

Parker, G. (2008) Parish and community-led planning, local empowerment and local evidence bases: an examination of 'best' practice, *Town Planning Review*, 79(1): 61–85.

Parker, G. and Amati, M. (2009) Institutional setting, politics and planning, *International Planning Studies*, 14(2): 141–60.

Parker, G. and Murayama, M. (2005) Doing the groundwork? transferring a UK environmental planning approach to Japan, *International Planning Studies*, 10(2): 123–140.

Parker, G. and Murray, C. (2012) Beyond tokenism? Community-led planning and rational choices: Findings from participants in local agenda-setting in England, *Town Planning Review*. 83(1): 1–28.

Parker, G. and Ravenscroft, N. (1999) Benevolence, nationalism and hegemony: fifty years of the National Parks and Access to the Countryside Act 1949, *Leisure Studies*, 18(4): 297–313.

Parker, G. and Wragg, A. (1999) Actors, networks and (de)stabilisation: the issue of navigation on the river Wye, *Journal of Environmental Planning and Management*, 42(4): 471–87.

Parsons, K. and Shuyler, D. (2003) From garden city to green city. The legacy of Ebeneezer Howard, *Journal of Planning Education and Research*, 23: 213–15.

Pearce, D., Hamilton, K. and Atkinson, G. (1996) Measuring sustainable development: progress on indicators, *Environment and Development Economics*, 1: 85–101.

Pennington, M. (2000) *Planning and the Political Market: Public Choice and the Politics of Government Failure*. London: Athlone Press.

Pennington, M. (2002) *Liberating the Land: The Case for Private Land-Use Planning*. London: IEA.

Petts, J. (ed.) (1999) *Handbook of Environmental Impact Assessment, vol. 1*. Oxford: John Wiley & Sons/Blackwell.

Plant, R. (1996) Citizenship, rights and socialism, in King, P.T. (ed.) *Socialism and the Common Good: New Fabian Essays*. London: Routledge.

Porter, M.E. (1990) *The Competitive Advantage of Cities*. Basingstoke: Palgrave Macmillan.

Porter, M.E. (1992) *Competitive Advantage: Creating and Sustaining Superior Performance*. Research Paper No. 10. London: PA Consulting.

Porter, M.E. (1998) *The Competitive Advantage of Nations*, New York: The Free Press.

Porter, M.E. (2003) The economic performance of regions, *Regional Studies*, 37: 549–78.

Potter, J. and Moore, B. (2000) UK Enterprise Zones and the attraction of inward investment, *Urban Studies,* 37(8): 1279–312.

Pressman, J. and Wildavsky, A. (1973) *Implementation: How Great Expectations in Washington are Dashed in Oakland; or, Why it's Amazing that Federal Programs Work at All*. Berkeley, CA: University of California Press.

Pretty, J. and Ward., H. (2001) Social capital and the environment, *World Development*, 29(2): 209–227.

Price Waterhouse Coopers (2007) *The Northern Way. Northern City Visions: A Review of City Region Development Programmes*, July 2007. Newcastle: Northern Way Secretariat.

Punter, J. and Carmona, M. (1997) *The Design Dimension in Planning: Theory Content and Best Practice*. London: Spon.

Putnam, R. (2000) *Bowling Alone: The Collapse and Revival of American Community*. New York: Simon and Schuster.

Raco, M., Parker, G. and Doak, J. (2006) Reshaping spaces of local governance. Community strategies and the modernisation of local government in England, *Environment and Planning C*, 24(4), 475–96.

Rasch, W. and Wolfe, C. (eds) (2000) *Observing Complexity. Systems Theory and Postmodernity*. Minneapolis: University of Minnesota Press.

Ratcliffe, J. (1974) *An Introduction to Town and Country Planning*. London: Hutchinson.

Ratcliffe, J. (1978) *An Introduction to Urban Land Administration*. London: Estate Gazette.

Raynsford, N. (2000) PPG3 – making it work, *Town and Country Planning*, September, 262–63.

Reade, E. (1969) Contradictions in planning, *Official Architecture and Planning*, 1179–1185.

Relph, E. (1976) *Place and Placelessness*. London: Pion.

Relph, E. (1981) *Rational Landscape and Humanistic Geography*. London: Croom Helm.

Rex, J. and Moore, R. (1967) *Race, Community and Conflict*. Oxford: Oxford University Press.

Roberts, M. and Greed, C. (eds) (2001) *Approaching Urban Design*. Harlow: Pearson.

Robertson, R. (1995) Glocalization: time-space and homogeneity-heterogeneity, in Featherstone, M., Lash, S. and Robertson, R. (eds), *Global Modernities*. London: Sage.

Rodgers, C. (2009) Property rights, land use and the rural environment, *Land Use Policy*, 26S: S134–S41.

Roseland, M. (2000) Sustainable community development: integrating environmental, economic and social objectives, *Progress in Planning*, 54(2): 73–132.

Rowley, A. (1994) Definitions of urban design: The nature and concerns of urban design, *Planning Practice and Research*, 9(3): 179–97.

RTPI (1994) *Planners as Managers: Shifting the Gaze*. London: Royal Town Planning Institute.

RTPI (2007) *GPN1: Effective Community Engagement and Consultation*. Revised July 2007. London: Royal Town Planning Institute.

RTPI (2010) *What Planning Does*. Online: http://www.rtpi.org.uk/what_planning_does/ (accessed 30 August 2010).

RTPI (2011) *Revised Learning Outcomes for RTPI Accredited Courses*. London: Royal Town Planning Insitute. Online: http://www.rtpi.org.uk/item/4514&ap=1 (accessed 20 December 2011)

Rugby Borough Council (2007) *Summary of Decision Notice: Application Number R07/0832/MAJP*, Rugby: Rugby Borough Council. Online: http://www.planningportal.rugby.gov.uk/decision.asp?AltRef=R07/0832/MAJP (accessed 15 December 2011).

Rydin, Y. (2003) *Urban and Environmental Planning in the UK*, 2nd edn. Basingstoke: Palgrave Macmillan.

Rydin, Y. (2010a) *The Purpose of Planning*. Bristol: Policy Press.

Rydin, Y. (2010b) *Governing for Sustainable Urban Development*. London: Earthscan.

Sabatier, P. (1986) Top-down and bottom-up approaches to implementation research: a critical analysis and suggested synthesis, *Journal of Public Policy*, 6(1): 21–48.

Sack, R. (1986) *Human Territoriality. Its Theory and History*. Cambridge: Cambridge University Press.

Sager, T. (1994) *Communicative Planning Theory*. Aldershot: Avebury.

Sanyal, B. (ed.) (2005) *Comparative Planning Cultures*. London: Routledge.

Saunders, P. (1983) *Urban Politics: A Sociological Interpretation*. London: Hutchinson.

Schön, D. (1983) *The Reflective Practitioner*. New York, NY: Basic Books.

Scott, A. (2001) Globalisation and the rise of city regions, *European Planning Studies*, 9(7): 813–26.

Scott, A. and Bullen, A. (2004) Special landscape areas: landscape conservation or confusion in the town and country planning system? *Town Planning Review*, 75(2): 205–30.

Scott, A., Agnew, J., Soja, E. and Storper, M. (2001) Global city-regions: an overview, in Scott, A. (ed.), *Global City-Regions: Trends, Theory, Policy*. Oxford: Oxford University Press.

Scott, W. and Gough, S. (2003) *Sustainable Development and Learning: Framing the Issues*. London: Taylor and Francis.

Sedjo, R. and Marland, G. (2003) Inter-trading permanent emissions credits and rented temporary carbon emissions offsets: some issues and alternatives, *Climate Policy*, 3(4): 435–44.

Seel, B., Paterson, M. and Doherty, B. (2000) *Direct Action in British Environmentalism*. London: Routledge.

Self, P. (1977) *Administrative Theories and Politics*, 2nd edn. London: Allen and Unwin.

Selman, P. (1999) *Environmental Planning*, 2nd edn. London: Paul Chapman.

Selman, P. (2000) Networks of knowledge and influence: connecting 'the planners' and 'the planned', *Town Planning Review*, 71(1): 109–21.

Selman, P. (2001) Social capital, sustainability and environmental planning, *Planning Theory and Practice*, 2(1): 13–30.

Selman, P. (2009) Conservation designations – are they fit for purpose? *Land Use Policy*, 25S: 142–153S.

Shaw. T. (2007) Editorial, *Journal of Environmental Planning and Management*, 50(5): 574–8.

Sheller, M. and Urry, J. (2006) The new mobilities paradigm, *Environment and Planning A*, 38(2): 207–26.

Shepherd, S., Timms, P. and May, A. (2006) Modelling requirements for local transport plans: an assessment of English experience, *Transport Studies*, 13(4): 307–17.

Shipley, R. (2002) Visioning in planning: is the practice based on sound theory? *Environment and Planning A*, 34: 7–22.

Silverman, D. (2000) *Doing Qualitative Research*. London: Sage.

Simmonds, R. and Hack, G. (Eds.) (2000) *Global City Regions: Their Emerging Forms*. London: Spon.

Simms, A., Kjell, P. and Potts, R. (2005) *Clone Town Britain*. London: New Economics Foundation.

Simon, H. (1997) *Models of Bounded Rationality. Empirically Grounded Economic Reason*, vol. 3. Cambridge, MA: MIT Press.

Skeffington, A. (1969) *People and Planning: Report of the Committee on Public Participation in Planning*. London: HMSO.

Smith, D. (1974) *Amenity and Urban Planning*. London: Granada.

Soja, E. (1996) *Thirdspace: Journeys to Los Angeles and Other Real and Imagined Places*. Oxford: Blackwell.

Soja, E. (2003) Writing the city spatially, *City*, 7(3): 269–80.

Sorensen, A. (2002) *The Making of Urban Japan*. London: Routledge.

Sorensen, A. (2010) Land, property rights, and planning in Japan: institutional design and institutional change in land management, *Planning Perspectives*, 25(3): 279–302.

Sowell, T. (1999) *The Quest For Cosmic Justice*. New York, NY: Free Press.

Sternberg, E. (1993) Justifying public intervention without market externalities: Karl Polanyi's theory of planning in capitalism, *Public Administration Review*, 53(2): 100–9.

Stevens, R. (2007) *Torts and Rights*. Oxford: Oxford University Press.

Stoker, G. (1998) Theory and urban politics, *International Political Science Review*, 19(2): 119–29.

Stone, C. (1989) *Regime Politics*. Lawrence, KA: University Press of Kansas.

Storper, M. (1997) *The Regional World: Territorial Development in a Global Economy*. New York, NY: Guilford Press.

Stubbs, M. (1997) The new Panacea? An evaluation of mediation as an effective method of dispute resolution in planning appeals, *International Planning Studies*, 2(3): 347–65.

Stubbs, M. (1997) The new panacea? An evaluation of mediation as an effective method of dispute resolution in planning appeals, *International Planning Studies*, 2(3): 347–65.

Svensson, G. (2001) 'Glocalization' of business activities: a 'glocal strategy' approach, *Management Decision*, 31(1): 6–18.

Swyngedouw, E. (2000) Authoritarian governance, power and the politics of rescaling, *Environment and Planning D: Society and Space*, 18: 63–76.

Tait, M. (2002) Room for manoeuvre? An actor-network study of central-local relations in development plan making, *Planning Theory and Practice*, 3(1): 69–85.

Tait, M. and Jenson, O. (2007) Travelling ideas, powers and place: the cases of urban villages and business improvement districts, *International Planning Studies*, 12(2): 107–28.

Taylor, N. (1998) *Urban Planning Theory*. London: Sage.

Tewdwr-Jones, M. (1999) Discretion, flexibility, and certainty in British planning: emerging ideological conflicts and inherent political tensions, *Journal of Planning Education and Research*, 18(3): 244–56.

Tewdwr-Jones, M. and Allmendinger, P. (1998) Deconstructing communicative rationality: a critique of Habermasian collaborative planning, *Environment and Planning A*, 30(11): 1975–90.

Tewdwr-Jones, M. and Allmendinger, P. (eds) (2002) *Planning Futures*. London: Routledge.

Tewdwr-Jones, M. and Phelps, N. (2000) Levelling the uneven playing field: inward investment, interregional rivalry and the planning system, *Regional Studies*, 34: 429–40.

Tewdwr-Jones, M., Morphet, J. and Allmendinger, P. (2006) The contested strategies of local governance: community strategies, development plans, and local government modernisation, *Environment and Planning A*, 38(3): 533–51.

Theory, Culture and Society (2005) Special issue on Complexity, *Theory, Culture and Society*, 22 (5): 1–274.

Thomas, H. and Healey, P. (ed.) (1991) *Dilemmas in Planning Practice: Ethics, Legitimacy and the Validation of Knowledge*. Aldershot: Avebury.

Thompson, G., Frances, J., Levacic, R. and Mitchell, J. (eds) (1991) *Markets, Hierarchies and Networks: The Coordination of Social Life*. London: Sage.

Thompson, N. (2005) Inter-institutional relations in the governance of England's national parks: a governmentality perspective, *Journal of Rural Studies*, 21(3): 323–34.

Thompson, N. and Ward, N. (2005) *Rural Areas and Regional Competitiveness*. Report to the Local Government Rural Network, October 2005. Centre for Rural Economy, University of Newcastle.

Thornley, A. (1991) *Urban Planning Under Thatcherism*. London: Routledge.

Tiesdell, S. and Adams, D. (2011) *Urban Design in the Real Estate Development Process*. Oxford: Wiley-Blackwell.

Tomaney, J. and Mawson, J. (eds) (2002) *England: The State of the Regions*. Bristol: Policy Press.

Tuan, Y.F. (1974) *Topophilia*. New Jersey, NJ: Prentice Hall.

Tuan, Y.F. (1977) *Space and Place: The Perspective of Experience*. Minneapolis, MN: University of Minnesota.

UNCED (1997) *Our Common Future* ('The Brundtland Report'). Oxford: Oxford University Press.

UNESCO (2010) *List of World Heritage Sites*. Online: http://whc.unesco.org/en/map/ (accessed 26 August 2010).

Untaru, S. (2002) Regulatory frameworks for place-based planning, *Urban Policy and Research*, 20(2): 169–86.

Urry, J. (1995) *Consuming Places*. London: Routledge.

Urry, J. (2000a) Mobile sociology, *British Journal of Sociology*, 51(1): 185–203.

Urry, J. (2000b) *Sociology Beyond Societies*. London: Routledge.

Urry, J. (2002) *Global Complexity*. Cambridge: Polity Press

Urry, J. (2007) *Mobilities*. Cambridge: Polity Press.
Ward, S. (ed.) (1992) *The Garden City: Past, Present and Future*. London: Routledge.
Ward, S. (1994) *Planning and Urban Change*. London: Paul Chapman.
Wates, N. (1990) *The Community Planning Handbook*. London: Department For International Development.
WCED (1987) *Our Common Future*. World Commission on Environment and Development. Oxford: Oxford University Press.
Webber, M. (1963) Order in diversity: community without propinquity, in Wingo, L. (ed.), *Cities and Space: The Future Use of Urban Land*. Baltimore: Johns Hopkins University Press, pp. 23–54.
Webster, C. (1998) Public choice, Pigouvian and Coasian planning theory, *Urban Studies*, 35(1): 53–75.
Webster, C. and Lai, L.W.C (2003) *Property Rights, Planning and Markets: Managing Spontaneous Cities*. Cheltenham: Edward Elgar.
West, P., Igoe, J. and Brockington, D. (2006) Parks and peoples: the social impact of protected areas, *Annual Review of Anthropology*, 35: 251–71.
While, A., Jonas, A. and Gibbs, D. (2004) The environment and the entrepreneurial city: searching for the urban 'sustainability fix' in Manchester and Leeds, *International Journal of Urban and Regional Research*, 28(3): 549–69.
Wilcox, D. (1994) *Guide to Effective Involvement*. Brighton: Partnership Books.
Wilkinson, S and Reed, R. (2008) *Property Development*, 5th edn. London: Spon.
Williams, R. (1977) *The Country and the City*. London: Chatto and Windus.
Williams, R. (1983) *Keywords*. London: Fontana.
West Yorkshire Local Transport Partnership. (2006) *West Yorkshire Local Transport Plan 2006–2011*. Online: http://www.wyltp.com/Archive/wyltp2/wyltp2006-11 (accessed 10 January 2012).
Woolcock, M. (1998) Social capital and economic development: toward a theoretical synthesis and policy framework, *Theory and Society,* 27: 151–208.
World Resources Institute (2008) *Ecosystem Services: A Guide for Decision Makers*. Washington, DC: World Resource Institute.
Yiftachel, O. (1998) Planning and social control: exploring the dark side, *Journal of Planning Literature*, 12(4): 395–406.

Useful Websites

ACSP – Association of Collegiate Schools of Planning: http://www.acsp.org/
AESOP – Association of European Schools of Planning: http://www.aesop-planning.com/
Planning Portal: http://www.planningportal.gov.uk/
CABE: http://www.cabe.org.uk/default.aspx?contentitemid=1436
Civic Trust: http://www.civictrust.org.uk/
DCLG – the Department for Communities and Local Government: http://www.communities.gov.uk/ This is the UK (England) government department responsible for planning.

Egan review: http://www.communities.gov.uk/pub/264/TheEganReviewSkillsforSustainableCommunities_id1502264.pdf The 2004 report on skills for sustainable communities.

English Heritage: http://www.english-heritage.org.uk/server/show/nav.1062 (see web pages on Conservation Areas)

Natural England, Landscape Character: http://www.countryside.gov.uk/LAR/Landscape/CC/countryside_character.asp

RTPI – Royal Town Planning Institute, the professional institute for town planning. The website includes a range of information about planning, planning education and practice: http://www.rtpi.org.uk/

RUDI – Resource for Urban Design Information: http://www.rudi.net/information_zone/design_guides_and_codes

TCPA – Town and Country Planning Association, a campaigning organisation aiming to inform planning practice: http://www.tcpa.org.uk

INDEX

Index

Index